FLORA OF TROPICAL EAST AFRICA

STERCULIACEAE

MARTIN CHEEK* AND LAURENCE DORR**

Monoecious or polygamous, evergreen or deciduous trees and shrubs, rarely herbs or climbers. Indumentum stellate, often mixed with simple hairs, rarely with peltate-lepidote scales. Stems sometimes exuding mucilage when wounded, fibrous. Leaves alternate, simple, elliptic to oblanceolate or ± orbicular and then often digitately lobed, venation palmate or pinnate, reticulate; petiole usually swollen and often kneed at base and apex. Stipules linear to lanceolate, usually caducous. Flowers actinomorphic, hypogynous, bisexual, perianth biseriate, calyx 5-lobed or of 5 free sepals (spathaceous in *Mansonia*) valvate. Corolla imbricate, with 5 free often clawed petals, or (genera 1–6) flowers unisexual, perianth uniseriate, the perianth 5-lobed, valvate. Androgynophore well-developed, inconspicuous or absent; androecium uniseriate, stamens (4–)5–20, all fertile or with some filiform staminodes, the filaments united into a long or short tube, or free or, (genera 1–6) anthers subsessile in a ± globose head on an androgynophore; less usually androecium biseriate, the inner whorl being staminodes, usually petaloid, the outer whorl fertile. Anthers dithecal, dehiscing by slits, rarely by apical pores, extrorse. Gynoecium syncarpous, locules (1–)5, placentation axile, ovules 2–numerous or apocarpous (genera 1–7), carpels 5(–numerous, then spiralled) cohering, separating in fruit; styles as many as carpels, cohering and appearing single, with 3–5 apical branches. Fruit syncarpous, then loculicidally, rarely septicidally (*Byttneria*) dehiscent, rarely indehiscent and then berry-like (*Theobroma*), if apocarpous, the fruitlets ventrally dehiscent along the placenta, then woody or leathery; or indehiscent and fleshy, leathery or papery. Seeds 1-numerous, rarely winged or arillate, endosperm present or absent, cotyledons flat or folded, thin or fleshy.

Mainly a tropical family but several species in the subtropics. About 67 genera and 1500 species.

The 'core' families of Malvales have traditionally been delimited as follows:

1. Stamens monothecal 2
   Stamens dithecal 3
2. Herbs and shrubs, very rarely trees, not spiny; pollen spiny ... *Malvaceae*
   Trees, often spiny; pollen usually smooth .... *Bombacaceae*
3. Stamens usually numerous, in multiple whorls; staminodes usually absent .... *Tiliaceae*
   Stamens 5–20 in a single whorl, often including staminodes, or in two whorls, the inner staminodal often petaloid, the outer all fertile .... *Sterculiaceae*

Cola, Dombeya, Hermannia, Octolobus, Pterygota and *Sterculia* by Martin Cheek
Byttneria, Harmsia, Heritiera, Hildegardia, Leptonychia, Mansonia, Melhania, Melochia, Waltheria and cultivated species by Laurence Dorr
Nesogordonia by Laurence Dorr and Lisa Barnett***

* Royal Botanic Gardens, Kew. [M. Cheek@kew.org]
** National Museum of Natural History, Dept. of Botany, MRC–166, Smithsonian Institution, P.O. Box 37012, Washington, DC 20013–7012, USA. [dorrl@ si.edu]
*** Smithsonian Tropical Research Institute, MRC–705, Smithsonian Institution, Washington, DC 20013–7012, USA.

The division of Sterculiaceae from Tiliaceae using this division has long been problematic, as illustrated by Hutchinson's provision of a key to the genera of the combined families (Hutchinson 1967, The Genera of Flowering Plants 2). Recent molecular work (e.g. Baum et al., Harvard Pap. Bot. 3: 315–330 (1998) and Bayer et al., Bot. J. Linn. Soc. 129: 267–303 (1999)) has shown that some groups of traditional Sterculiaceae are undoubtedly more closely related to some parts of traditional Tiliaceae than they are to the rest of the family. While Malvaceae and Bombacaceae can be maintained ± in the traditional sense (although *Durio* and its SE Asian relatives require removal from the last), there is no doubt that this is not feasible regarding Tiliaceae and Sterculiaceae, which need to be broken up into several units, all of which are morphologically distinct and have previously been recognised as families. Core Malvales can thus be expanded from the traditional four to ten newly circumscribed families. So far as FTEA is concerned these are as follows:

> Brownlowiaceae (*Christiana* and *Carpodiptera*, earlier published in FTEA Tiliaceae, 2001)
> Sparrmanniaceae (the rest of the genera in FTEA Tiliaceae, 2001)
> Bombacaceae (as published in FTEA Bombacaceae, 1989)
> Malvaceae (in preparation for FTEA)
> Sterculiaceae (genera 1–6 of this volume)
> Helicteraceae (genus 7)
> Pentapetaceae (genera 8–11)
> Byttneriaceae (genera 12–16)

An alternative approach is to unite all these families in a 'Super-Malvaceae' and to recognise them at the subfamily level instead (Bayer & Kubitzki, Genera and Families of Vascular Plants 5 (2001)). This seems the less desirable solution since it treats Malvaceae as a dustbin family and creates greater taxonomic instability at the family level than the solution above.

Each species in this account has been assessed for its conservation status by Martin Cheek using the criteria of IUCN (2001). IUCN Red List Categories: Version 3.1. IUCN Species Survival Commission, IUCN, Gland, Switzerland and Cambridge, UK.

Both authors thank Quentin Luke for painstakingly checking the manuscript while in proof, and providing many new records that extended the previously known altitudinal and geographical ranges of the taxa. These additions were based on his own prolific collections and those at the East African herbarium at Nairobi.

KEY TO THE GENERA

1. Perianth uniseriate, (4–)5(–8)-lobed; male flowers with androgynophore; fruits apocarpous; usually trees 2
   Perianth biseriate (with distinct petals and sepals/calyx), petals free; flowers hermaphrodite; fruits syncarpous (apocarpous in *Mansonia*), often shrubs, less usually (e.g. *Nesogordonia, Mansonia, Dombeya*) trees 7
2. Perianth 8-lobed; carpels 10+, spiralled . . . .   3. *Octolobus* (p. 46)
   Perianth 5(–6)-lobed (but 8-lobed in *Cola ruawensis* and *C. octoloboides*); carpels 5, uniseriate 3
3. Fruits dry when mature; dispersal units (whether fruitlets or seeds) dry, winged 4
   Fruits fleshy or leathery when mature; dispersal units (seeds) ellipsoid, not winged 6

4. Fruitlets ± strongly winged, elongate, 1-seeded,
    indehiscent 5
    Fruitlets not winged, globose, dehiscing by 2 valves,
    with 20+ winged seeds . . . .   *4. Pterygota* (p. 48)
5. Leaves pinnately nerved, lower surface covered in
    fimbriate-peltate scales; seeds filling entire fruitlet
    cavity; fruitlet leathery, glossy; mangrove . . . .   *5. Heritiera* (p. 52)
    Leaves palmately nerved, lower surface glabrous,
    fruit cavity mostly air-filled; fruitlet papery, matt;
    coastal thicket . . . .   *6. Hildegardia* (p. 54)
6. Usually small trees of undisturbed evergreen forest;
    fruitlets indehiscent, placenta glabrous; seeds
    filling the fruitlet, often faceted by mutual
    compression, radicle pointing towards hilum . . .   *2. Cola* (p. 20)
    Usually trees or shrubs of wooded grassland or
    disturbed or deciduous forest; fruitlets dehiscent,
    placenta with irritant hairs; seeds not filling the
    fruitlet, never faceted, radicle pointing away from
    hilum . . . .   *1. Sterculia* (p. 5)
7. Trees 8
    Subshrubs or shrubs, rarely annual herbs or climbers 11
8. Leaves pinnately nerved 9
    Leaves digitately/palmately nerved 10
9. Fruit 5-angled, loculicidal, seeds winged . . . .   *11. Nesogordonia* (p. 92)
    Fruit ellipsoid, indehiscent, seeds unwinged;
    cultivated, sometimes naturalised . . . .   *Theobroma*
10. Fruits apocarpous, fruitlets winged, indehiscent;
    flowers with androgynophore; rare, thicket, **T** 3 . .   *7. Mansonia* (p. 57)
    Fruits syncarpous, ± ellipsoid, dehiscent; flowers
    without androgynophore; widespread and common 11
11. Leaves with gland at base of midrib below . . . .   *12. Byttneria* (p. 95)
    Leaves without gland or glands on midrib 12
12. Leaves usually oblong-elliptic, at least twice as long
    as broad 13
    Leaves ± orbicular in outline, ± as long as broad . .   *8. Dombeya* (p. 59)
13. Shrubs and climbers of evergreen forest 14
    Trees, shrubs, subshrubs and herbs of thicket,
    bushland and grassland 16
14. Leaf margin toothed (except some *D. shupangae*);
    petals much longer than sepals, not cupped at
    base . . . .   *8. Dombeya* (p. 59)
    Leaf margin entire (except some juveniles); petals
    either much shorter than sepals OR cupped at
    base 15
15. Petals with cup, terminal appendage and basal claw;
    fertile stamens 5, each opposing and concealed
    by a petal cup; fruits spiny . . . .   *12. Byttneria* (p. 95)
    Petals subelliptic, lacking cup, claw or appendage;
    fertile stamens in 5 phalanges mixed with
    staminodes; fruits smooth . . . .   *13. Leptonychia* (p. 99)
16. Epicalyx of three bracts, often conspicuous and
    persisting in fruit (but caducous in most *Dombeya*);
    androecium in a single whorl of 15–20 stamens;
    fertile stamens mixed with longer staminodes 17
    Epicalyx usually absent; androecium of 5 stamens;
    all stamens fertile 19

17. Shrubs or subshrubs usually less than 0.5 m(–2 m)
   tall; petals yellow; fruits spiny or ± concealed
   within the epicalyx 18
   Trees or shrubs usually more than 2.5 m tall when in
   flower; petals white or pink; fruits smooth, naked   8. *Dombeya* (p. 59)
18. One fertile stamen alternating with each staminode;
   fruit unarmed, usually ± concealed in epicalyx . . .   9. *Melhania* (p. 73)
   Three fertile stamens alternating with each staminode;
   fruit spiny, naked . . . .   10. *Harmsia* (p. 90)
19. Anthers with lines of conspicuous bristles, filaments
   free or mostly free; fruits with 2–20 reniform seeds
   per locule . . . .   14. *Hermannia* (p. 104)
   Anthers glabrous, filaments largely united into a
   cylindrical tube; fruits with 1(–2) trigonal or
   obovoid seeds per locule 20
20. Ovary and fruit 1-locular, seeds obovoid . . . .   16. *Waltheria* (p. 127)
   Ovary and fruit 5-locular, seeds trigonal . . . .   15. *Melochia* (p. 122)

CULTIVATED TAXA

Abroma augustum (*L.*) *J.A.Murray*, Syst. Veg., ed. 14, 696 (1784) as *angusta*;
T.T.C.L.: 592 (1949); Germain in F.C.B. 10: 211 (1963), as *augusta*
   Shrub to 6 ft tall. Lower leaves to 12.5 × 13 cm, 3–5-lobed, upper leaves simple,
ovate. All parts of plant are irritating. "Devil's Cotton"; native to tropical Asia.
TANZANIA. East African Agricultural Research Station Amani, 27 Aug. 1929, *Greenway* 1706!

Brachychiton acerifolius (*G. Don*) *F. Muell.*, Fragm. 1: 1 (1858); I.T.U.: 13 (1953);
Wild in F.Z. 1(2): 557 (1961); Wild & Gonç. in F.M. 27: 46 (1979), as *acerifolium*; Noad
& Birnie, Trees Kenya: 271, fig. p. 272 & t. 18 (1989). Syn.: *Sterculia acerifolia* G. Don,
Gen. Hist. Dichl. Pl. 1: 517 (1831); U.O.P.Z.: 454 (1949)
   "Flame Tree", "Australian Flame Tree". Native to Australia, where it is found along
the east coast. This species frequently is cultivated as an ornamental, such as in
Uganda and on Zanzibar, Victoria Garden.

Brachychiton populneus (*Schott & Endl.*) *R.Br.* in Horsfield, Pl. Jav. Rar. 234 (1844);
I.T.U: 13 (1953); Wild in F.Z. 1(2): 557 (1961); Wild & Gonç. in F.M. 27: 46 (1979),
as *populneum*; Noad & Birnie, Trees Kenya: 271, fig. p. 272 (1989); Thulin, Fl. Som.
2: 37 (1999). Syn.: *Poecilodermis populnea* Schott & Endl., Melet. Bot. 33 (1832).
   "Australian Bottle Tree", "Bottle Tree", "Kurrajong". East African material can be
referred to *B. populneus* subsp. *populneus*, which is native to eastern and southeastern
Australia but now widely cultivated as an ornamental.
UGANDA. (fide Dale, 1953)
KENYA. Nairobi, cult. Denis Pritt road, 21 Feb. 1968, *Gillett* 18531! and Mua Lane, Muthega, 24
   Sep. 1965, *Greenway* 11928!
TANZANIA. Cult. tree in garden at Chala Mission, Ufipa Plateau, 1949–51, *Bullock* s.n.!

Firmiana simplex (*L.*) *Wight*, U.S. Dept. Agric. Bur. Pl. Industry, Bull. 142: 67
(1909). Syn.: *Firmiana plantanifolia* (L.f.) Schott & Endl., Melet. Bot. 33 (1832); I.T.U:
42 (1953); *Sterculia plantanifolia* L.f., Suppl. Pl. 423 (1781)
   "Japanese Varnish Tree".
UGANDA. Cultivated in Entebbe Gardens.

Fremontodendron californicum (Torr.) Colville × **F. mexicanum** *Davidson*
("California Glory")
   Native to California (U.S.A.) and adjacent Mexico. The large, showy, spreading-
campanulate, yellow, petal-like calyx that is leaf-opposed, and the 5- or 4-valved, densely
setose capsules readily distinguish *Fremontodendron* from the native Sterculiaceae with

which family it is traditionally placed although recent molecular work (see introduction above) has placed it in Bombacaceae.

KENYA. Mau Narok, 23 June 1970 (fl, fr), *Gillett* 19222! This specimen, which is heavily insect-damaged, was identified as *F. californicum* (Torr.) Colville, but probably is the hybrid.

Guazuma ulmifolia *Lam.*, Encycl. 3: 52 (1789); Wild in F.Z. 1(2): 517 (1961); Wild & Gonç. in F.M. 27: 57 (1979)

Native to tropical and subtropical America; occasionally cultivated in the Old World tropics.

KENYA. Arboretum, Nairobi, 5400 ft, May 1941 (fl), *Mrs. Joy Bally in P.R.O. Bally* 1466 (K!)

Kleinhovia hospita *L.*, Sp. Pl., ed. 2: 1365 (1763); Mast. in F.T.A. 1: 226 (1868); I.T.U.: 47 (1953).

Native to the Old World tropics and subtropics.

UGANDA. Jinja, Entebbe (fide Dale, 1953)

Pterospermum acerifolium *Willd.*, Sp. Pl., ed. 4, 3: 729 (1800); U.O.P.Z.: 427 (1949); Guide E. Afr. Fl. Trees: 12 (1968)

Native to India. Shrub or tree.

KENYA. Arboretum, Nairobi, Oct 1941, *Bally* 1463 (K!); Kenyatta College, ornamental, 19 Mar 1980, *Gillett* 22799!; Also on Zanzibar.

Theobroma cacao *L.*, Sp. Pl. 782 (1753); T.T.C.L.: 604 (1949); U.O.P.Z.: 467, fig. (1949); I.T.U.: 68 (1953); Germain in F.C.B. 10: 211 (1963); Vollesen in Fl. Eth. 2(2): 165 (1995)

Cocoa. An important crop in West Africa; source of chocolate. Sometimes collected in secondary forest and thought to be wild in our area.

UGANDA. Entebbe Botanic Garden

KENYA. Near Kwale (Luke, pers. comm.).

TANZANIA. Cultivated at Amani, on Pemba, and Zanzibar. Tanga: Muheza District: W-facing slopes of Kwamgumi Forest Reserve, 12 June 2000, *Mwangoka & Maingo* 1358!; Morogoro District: Kasanga Forest Reserve, 26 July 2000, *Mhoro* UMBCP 167!

1. **STERCULIA**

L., Sp. Pl.: 1008 (1753) & Gen. Pl., ed. 5: 438 (1754)

Monoecious, evergreen or deciduous shrubs and trees; stem often producing a gummy exudate when wounded; bud-scales present and persistent. Leaves entire, digitately lobed or divided; lobes or leaflets 3–7, entire; stipules usually highly caducous. Inflorescences usually paniculate, rarely spike-like, often on leafless stems; bracts usually caducous. Flowers with a single united perianth whorl, purplish red, green or yellow, usually with five reflexed lobes, rarely cohering at the tip. Male flowers with 8–10 anthers each with two separate thecae, ± randomly dispersed (not in discrete ranks as in *Cola*) in a globose head, ovary entirely absent, borne on a long, often curved androphore. Female flowers with anthers at base of subsessile apocarpous ovary; carpels 5, uniseriate. Fruit carpels separated, developing into dehiscent follicles, each globose to cylindrical, often rostrate and stipitate, pericarp leathery and flexible to stout and woody, the valves opening slightly or completely, outside often red or pink, inside pale yellow or white, dehiscing along the line of the placenta which usually covered in urticating hairs. Seeds leaden black, often with a yellow or red aril near the seed stalk. Fig. 1, p. 6.

About 150 species throughout the tropics, 27 in Africa and Madagascar.

In East Africa the bark of many species is used locally as a source of fibre for ropes and string, and the seeds when roasted are often used as food.

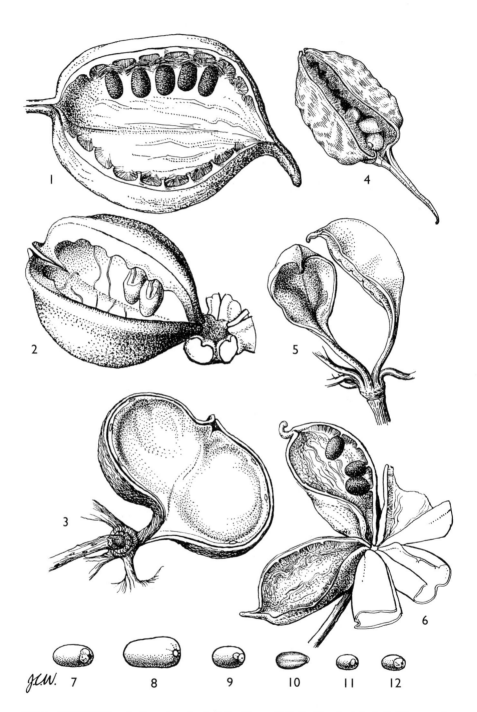

FIG. 1. *STERCULIA* — fruits and seeds. **1**, **7**, *S. africana*. **2**, **8**, *S. appendiculata*. **3**, **9**, *S.tragacantha*. **4**, **10**, *S. rhynchocarpa*. **5**, **11**, *S. quinqueloba*. **6**, **12**, *S. mhosya*. All illustrations × 1. Reproduced from F.Z. 1: t. 104. Drawn by Joanna Webb, with 4 & 10 added by Hazel Wilks.

1. Leaf-blade simple, pinnately veined 2
   Leaf-blade digitately lobed or divided, or if simple
   (rarely), digitately veined 5
2. Leaf-blade greenish white beneath, completely
   covered in white scales with fimbriate margins   1. *S. subviolacea* (p. 8)
   Leaf-blade never greenish white beneath beneath,
   but green or brown, covered in brown stellate
   hairs, without white scales 3
3. Leaf-blade tomentose beneath; perianth lobes
   cohering at apex. **T** 4 . . . .   2. *S. tragacantha* (p. 9)
   Leaf-blade tomentellous to glabrescent beneath;
   perianth lobes not cohering at the apex, but
   often reflexed 4
4. Leaf-blade cordate at base, lower surface
   glabrescent, bearing only scattered cruciform
   hairs; flowering with the leaves. **U** 2, 4; **K** 5 . . .   3. *S. dawei* (p. 10)
   Leaf-blade not or rarely cordate at the base, lower
   surface tomentellous to glabrescent with 5–8-
   armed stellate hairs; flowering when leafless.
   K 7; **T** 6,8 . . . .   4. *S. schliebenii* (p. 10)
5. Leaf-blade completely digitately divided into 5–7
   leaflets (cultivated) . . . .   *S. foetida* (p. 20)
   Leaf-blade digitately lobed or divided for up to $\frac{1}{2}$
   its length, rarely entire 6
6. Trunk pale grey, greenish yellow or white; leaf-
   blade usually 7-lobed, 10–24 × 7–19 cm, glabrous
   beneath; seeds white to pale yellow, suspended
   from the thick-walled fruit on white threads
   1–2 cm long . . . .   5. *S. appendiculata* (p. 11)
   Trunk grey, brown or mottled; leaf-blade usually
   3–5-lobed (sometimes 7-lobed in *S. mhosya*) or
   entire, less than 10 × 7.2 cm if glabrous
   beneath; seeds black, not suspended from the
   fruits on white threads, but firmly fixed within 7
7. Leaves tomentose below 8
   Leaves thinly puberulent, subscabrid or glabrescent
   below 12
8. Bole flaking in large plates, so variegated red and
   white; leaves 8–36 × 10–34 cm . . . .   6. *S. quinqueloba* (p. 12)
   Bole uniformly grey or brown; leaves to 17 × 17 cm,
   usually smaller 9
9. Stipules persistent, conspicuous; fruits long-
   cylindrical and moniliform, pericarp leathery
   when mature . . . .   9. *S. stenocarpa* (p. 16)
   Stipules caducous, rarely seen; fruits ovoid or
   ellipsoid, pericarp thickly woody when mature 10
10. Leaves usually < 5 cm wide; fruits grey-green,
   subtomentellous, with sharply raised longitudinal
   ridges, distinctly prickly and with a rostrum
   (1–)2–4.5 cm long . . . .   10. *S. rhynchocarpa* (p. 17)
   Leaves usually > 5cm wide; fruits yellowish-brown,
   tomentose, without surface sculpture, rostrum
   0–2 cm long 11

11. Leaves 3–5-lobed, 8.5–17.5 cm long and broad,
    densely yellow-brown tomentose below; outer
    perianth tube completely obscured by hairs ..   7. *S. setigera* (p. 14)
    Leaves entire or 3-lobed, 3.5–10(–12.5) cm long,
    (3–)5–8(–11) cm wide, tomentose to glabrescent
    below; outer perianth tube not completely
    obscured by hairs ....   11. *S. africana* (p. 18)
12. Leaf-blade usually divided for ½, lobes acuminate;
    inflorescence with purplish glandular hairs ...   8. *S. mhosya* (p. 15)
    Leaf-blade usually divided for ½ or less, lobes usually
    rounded; inflorescence never with hairs purplish
    or glandular 13
13. Leaf-blade (0.5–)1–5(–9) × (0.5–)0.8–5(–8.5) cm;
    fruits grey-green, subtomentellous, with sharply
    raised longitudinal ridges, distinctly prickly ...   10. *S. rhynchocarpa* (p. 17)
    Leaf-blade (3.5–)4–10(–12.5) × (3.2–)5–8(–12) cm;
    fruits yellowish-brown, tomentose, without
    surface sculpture ....   11. *S. africana* (p. 18)

1. **Sterculia subviolacea** *K.Schum.* in P.O.A. C: 271 (1895); T.T.C.L.: 601 (1949); Wild in F.Z. 1: 556 (1961); Germain in F.C.B. 10: 272, t. 24 (1963). Type: Tanzania, Mwanza District: E Uzinza [Usindscha], *Stuhlmann* 3549 (B†, holo.); Mwanza District: Geita, *Carmichael* 473 (K!, neo.; EA!, isoneo. fide Wild 1961)

Evergreen tree 12–27 m tall; bole sometimes fluted or buttressed; ultimate branchlets 4–4.5 mm thick, pale greyish-brown, ± glabrous, drying with numerous fine longitudinal creases; bud-scales 4–6, oblong, acuminate, 3.5–4.5 mm long, 1–3 mm wide, caducous. Leaf-blade elliptic to oblong or slightly obovate, 12–23 cm long, 7–14 cm wide, apex rounded, rarely obtuse or subacuminate, base rounded to truncate, rarely slightly subcordate, very leathery, glabrous and drying pale green above, completely covered by a greenish white layer of interlocking, individually inconspicuous fimbriate peltate scales beneath; midrib, secondary and tertiary veins prominent below, covered with a fragile waxy layer; petiole terete, 15–40 mm long, 1.5–2.5 mm wide, sometimes swollen at base and tip, very sparsely stellate-hairy at first, soon becoming glabrous; stipules caducous. Inflorescence borne with leaves, 1–4 per stem from the highest leaf axil, 9–20 cm long, covered in long, woolly purplish-red stellate hairs, lowest branch 2–3.5 cm from the base, 2–5 cm long, bearing 4–7 partial peduncles, each (1–)3-flowered; bracts caducous, 3.5–5 mm long, 1–2 mm wide, indumentum as peduncle; pedicels 4–5 mm long. Flowers reddish–purple; perianth campanulate, 5–7.5 mm long, 5–6.5 mm wide, divided halfway into 5 lobes initially joined at the tips. Male flowers with androphore 1.5–1.8 mm long. Female flower with gynophore 0.5 mm long, with ± sessile, reduced stamens at the base of the globose, tomentose ovary. Fruits with undehisced follicles scarlet, 6.5–8 cm long, 2.5 cm wide; when dehisced becoming wider than long, wall 4–5 mm thick, rostrum 6–9 mm long, stipe 0.8–1.2 cm long. Outer surface reddish brown with longitudinal wrinkles, subscabrid with appressed stellate hairs, inner surface softly tomentose, hairs brown; seeds not known, but without a peg-like stalk persisting on the fruit.

TANZANIA. Mwanza District: Lake Victoria, Kiagamba, Nov. 1954, *Carmichael* 473!; Buha District: Kasulu, Nkotagu, Dec. 1954, *Forcus* 10!; Kigoma District: 5 km Uvinza/Kasulu crossroad–Kigoma, July 1951, *Eggeling* 6182!
DISTR. **T** 1, 4; Cameroon, Congo-Kinshasa, Angola, Zambia
HAB. Gallery and swamp forest; 900–1500 m

SYN. *S. ambacensis* Hiern in Cat. Welw. Afr. Pl. 1: 83 (1896); K. Schum. in E.M. 5: 104 (1900). Type: Angola, Puricacarambola, *Welwitsch* 4695 (BM!, holo.)

S. katangensis De Wildemann in Ann. Mus. Congo, sér. IV, 1: 211 (1903). Type: Congo-Kinshasa, Lufaku, *Verdick* 383 (BR, holo.)

LOCAL USES. None recorded

CONSERVATION NOTES. This species is here assessed as "Least Concern" in view of its large geographic range and because its habitat is ± undisturbed through most of its range, so far as is known.

NOTES. The white lower leaf surface and purplish, woolly hairy inflorescence are each highly characteristic for *S. subviolacea* amongst the species in East Africa with entire leaf-blades. Only seven specimens have been seen from East Africa. Fruiting collections with seeds are needed.

2. **Sterculia tragacantha** *Lindley* in Bot. Reg. 16: t. 1353 (1830); Mast. in F.T.A. 1: 216 (1868); K. Schum. in E.M. 5: 102, t. 9/F (1900); Wild in F.Z. 1: 556, t. 104/C (1961); Germain in F.C.B. 10: 269 (1963); Troupin in Flore Pl. Lign. Rwanda: 667, t. 230/1 (1982). Type: Sierra Leone, cult. England, Bot. Reg. 16, t. 1353 (1830)

Tree 3–24 m tall; bole cylindrical, sometimes with fluted buttresses; bark rough, grey or greyish brown, often deeply fissured; slash pale pink or pale orange, quickly becoming deep orange, exudate clear; ultimate branchlets rough and grey, 5–10 mm thick, the apex rusty tomentose. Leaf-blade elliptic-oblong or slightly obovate, (6.5–)11–21(–30) cm long, (4.7–)6.5–13(–16) cm wide, apex rounded to acuminate, base truncate, rounded, or subcordate, texture papery or leathery, glabrous, sometimes glossy above, tomentose (sometimes tomentellous in shade or on sucker growth) with red-brown 5–6-armed stellate hairs beneath; petiole terete, slightly swollen at base and tip, (2.5–)5–7.5 cm long, 1–2 mm wide, tomentellous with pale brown stellate hairs; stipules caducous. Inflorescences borne with the leaves, 2–10 per stem, each inflorescence 10–14(–22) cm long, 2.5–5(–8) cm wide, 2–3 mm thick at the base, with 20–25(–30) branches; pedicels 1–4.5 mm long. Flowers with perianth campanulate, rose-pink, pinkish or brownish purple, rarely green tinged with red, 4–7(–8) mm long, 3–5 mm wide, divided into 5 acute lobes 3–4 mm long, 1.5 mm wide, cohering at the apex, margins reflexed, outer surface with a mixture of short, wide, flaccid, ± appressed, colourless hairs and fine, pale brown stellate hairs; inner surface glabrous apart from the extreme base and the lobes which densely covered in long, stout pointed, patent white or purple hairs. Fruit with 3–5 follicles, follicles ellipsoid, 5.7–10 cm long, 2.5 cm wide, dehiscing flat, then 3.6–8.9 cm long, 4–8 cm wide, rostrum rather short, blunt, 3–5(–10) mm long, stipe stout, (0.5–)1(–2) cm long, pericarp thick, woody, 2–3 mm thick, outer surface red, then rusty brown, tomentellous, inner surface yellow-brown, stiffly and thickly tomentose; seeds glossy black, ellipsoid, 10–13(–18) mm long, 7–9(–12) mm wide, hilum white, terminal, elliptic, 1–2 mm long, with a small, globose, orange aril 0.5–1 mm wide on a short margin; seeds sessile, leaving a white scar on the pericarp wall. Fig. 1/3, 10, p. 6.

TANZANIA. Buha District: 50 km S of Kibondo, Mukugwe R., July 1951, *Eggeling* 6206!; Kigoma District: Mzuma, 4 Aug. 1958, *Mahinde* HSM/203!; Mpanda District: Kungwe-Mahali Peninsula, 1 Sep. 1959, *Harley* 9497!

DISTR. **T** 4; Guinea, Sierra Leone, Liberia, Ivory Coast, Ghana, Benin, Nigeria, Cameroon, Equatorial Guinea, Central African Republic, Congo-Kinshasa, Rwanda, Angola, Zambia

HAB. Swamp and riverine forest; 750–1700 m

LOCAL USES. None recorded.

CONSERVATION NOTES. This species is here assessed as "Least Concern" in view of its large geographic range. It is a fast-growing pioneer species.

NOTES. This plant has been reported, presumably on account of its specific epithet, to be a source of gum tragacanth. However, that production is derived only from species of *Astragalus* (Leguminosae). Whilst the gum from *Sterculia tragacantha* may resemble that of gum tragacanth as Lindley (Bot. Reg. 16: t.1353 (1830)) has mentioned, no reference has been found to its use as a substitute.

Sterculia tragacantha is closely related to *S. schliebenii* and *S. dawei*, q.v. for diagnostic characters.

3. **Sterculia dawei** *Sprague* in Bull. Herb. Boiss. sér. 2, 5: 1167 (1905); I.T.U.: 422 (1952); Hamilton, Uganda For. Trees: 118 (1981); K.T.S.L.: 167 (1994). Types: Uganda, "all over", *Dawe* 143; Uganda, Mengo District: Busiro, *Dawe* 206 (both K!, syn.)

Tree 9–27 m tall, to 1.5 m DBH; bole cylindrical, lacking buttresses or with large buttresses to 2.5 m, bark rough, brown to grey, flaking off in small pieces ± 1 cm wide; slash pink or red, with a reticulate pattern, rapidly turning brown; ultimate branchlets grey, rough and wrinkled, 5–8 mm wide; bud-scales triangular, 5 mm long, 2 mm wide, densely tomentose, with pale brown patent stellate hairs. Leaf-blade widely elliptic, rarely broadly ovate or suborbicular, (10–)11–16 cm long, (5.5–)10–16(–19) cm wide, apex rounded to subacute, sometimes shortly subacuminate, base cordate, the sinus (0.3–)1–1.5(–3) cm deep, texture papery, glabrous and drying dark green above, subscabrid with sparse, pale brown cruciform hairs beneath, drying grey; petiole slightly swollen at base and tip, terete, (2–)4–6.5 cm long, 1–3 mm wide, glabrous, rarely subscabrid, with indumentum as lower leaf surface; stipules caducous. Inflorescences produced with the leaves, (1–)2–7 per stem, each 5–15.5 cm long, 1.5–4.5 mm wide with 10–35 branches, indumentum as bud-scales; bracts ovate-acuminate, 7 mm long, 3 mm wide, pale brown tomentose; lowest branch 2–4 cm from the base, 0.8–3 cm long, bearing 1–4 partial peduncles each 5–9-flowered; pedicels 1.5–4.5 mm long. Flowers green outside, deep red inside, with campanulate perianth 4–7 mm long, 4–5 mm wide, divided into 5 triangular-acute lobes 1–1.5 mm long, 0.7–1 mm wide, outside with appressed, wide, flaccid, colourless, simple hairs, inner surface largely glabrous except the teeth and extreme base which densely covered in fine white simple hairs. Fruit usually with 3 follicles; follicles ellipsoid, 6–10 cm long, 3–4.5 cm wide, dehiscing flat, then ± 4.5 cm long, 8–10.5 cm wide, rostrum short and blunt, 0.3–0.9 cm long, stipe stout, 0.7–1 cm long, pericarp woody, 3–6 mm thick, outer surface red, then rusty brown, tomentellous, inner surface softly tomentose, pale brown; seeds glossy black, ellipsoid, 13–19 mm long, 9–11 mm wide, hilum white, terminal, elliptic 1–2 mm long with on a long margin a small globose orange aril, 0.5–1 mm wide; seed-stalk absent, detached seed leaving a white elliptic scar on the inner pericarp surface.

Uganda. Toro District: Bwamba, Kitengya, Aug. 1937, *Eggeling* 3365!; Mengo District: Entebbe, Dec. 1931, *Eggeling* 132! & Busiro, *Dawe* 206!
Kenya. North Kavirondo District: Marach, 14 Mar. 1960, *Coffee Officer* (*Elgon Nyanza*) H86/60!
Distr. U 2, 4; **K** 5; Cameroon, Gabon, Congo-Kinshasa, Angola
Hab. Forest; 900–1500 m

Syn. *S. bequaertii* De Wildeman in Ann. Soc. Sc. Brux. 40 (1): 188 (1921); Hallé in Fl. Gabon 2: 17, t. 3 (1961); Germain in F.C.B. 10: 271 (1963). Type: Congo-Kinshasa, Ituri, *Bequaert* 2384 (BR, holo.), **syn. nov.**
   S. tragacantha Lindl. var. *cruciata* Vermoesen in Man. Ess. For. Congo Belge: 258 (1923). No type indicated.
   S. purpurea Exell in J. B. 65, suppl. 1: 37 (1927). Type: Angola, Mayombe, *Gossweiler* 5393, 5402 (both BM, syn.), **syn. nov.**

Local uses. None recorded
Conservation notes. This species is here assessed as "Least Concern" in view of its large geographic range and because of its common habitat.
Notes. The leaves of *S. dawei* are similar to and so confused with those of *Chlorophora* (*Milicia*), Uganda's most valuable timber tree. In East Africa, *Sterculia dawei* is closely related to *S. schliebenii* and *S. tragacantha*. It is most easily distinguished from both by its cordate leaves, with characteristic cruciform hairs sparsely distributed beneath. It is odd that *S. purpurea* and *S. bequaertii* have not been reduced to synonymy before now.

4. **Sterculia schliebenii** *Mildbr.* in N.B.G.B. 12: 519 (1935); T.T.C.L.: 604 (1949); Wild in F.Z. 1: 557 (1961); K.T.S.L.: 168 (1994). Type: Tanzania, Lindi District: Lake Lutamba, 40 km W of Lindi, *Schlieben* 5243 (B, holo., probably destroyed; lecto. MAD, designated by Dorr in K.B. 59: 161 (2004); LISC, P!, S, iso.)

Deciduous tree 4–16 m tall, sometimes branching from the base; bark very rough, grey or brown, slash pink, watery; ultimate branchlets ridged or smooth, 3–5(–8) mm wide, becoming grey with scattered tuberculate lenticels; bud-scales narrowly triangular, 4–5 mm long, 1–2 mm wide. Leaf-blade obovate, sometimes slightly pandurate, (5–)8–18(–22) cm long, (3–)6–10(–1) cm wide, apex rounded to truncate, subacuminate, base obtuse, rarely truncate or subcordate, papery, glabrous to subscabrid above, tomentellous, subscabrid to glabrescent beneath with small fine pale brown appressed stellate hairs; petiole terete, (1.5–)3–5.5(–8) mm long, 1–2 mm wide, pale brown, subscabrid with indumentum as leaf-blade; stipules caducous. Inflorescences in clusters of 3–13 from the apex of leafless stems, each 4–7 cm long, 1–1.5(–3) cm wide, indumentum of purple, patent hairs, main axis bearing 9–12 short branches, the lowest 0.5–2 cm from the base, 0.3–2.1 cm long, bearing 1–3 partial-peduncles each with 1–3(–5) flowers; bracts narrowly triangular to linear, 4–5 mm long, 1–2 mm wide; pedicels 1–5 mm long. Flower perianth widely campanulate, 3–3.5(–7) mm long, 4–4.5(–8.5) mm wide, divided into 4–5(–7) triangular, acute, reflexed lobes 2–2.5 mm long, 2 mm wide, the outer surface pink, drying red, with purplish patent stellate hairs, the inner surface greenish white, drying grey, with short white simple hairs. Male flowers with androphore 2–3 mm long. Female flowers with gynophore slightly tapered, 0.5–1 mm long, glabrous, sometimes slightly angular; anthers reduced, ± sessile at ovary base; ovary globose, 2–2.3 mm wide, densely grey-brown tomentose, style 0.5 mm long. Fruit known only from one old, fallen, seedless follicle, elliptic, 9.5 mm long, 7.5 mm wide, 1.5 mm thick, inner surface tomentose, stellate; seed unknown.

KENYA. Kwale District: Shimba Hills, Mwele Mdogo Forest, 6 Jan. 1988, *Luke* 896! & Shimba Hills, SW Pengo Hill, 8 Nov. 1970, *Faden, Evans & Mahasi* 70/826! & Marenji Forest Reserve, 25 Feb. 1989, *Luke & Robertson* 1739!
TANZANIA. Uzaramo District: Pugu Hills, 29 Aug. 1982, *Hawthorne* 1683!; Lindi District: Rondo Plateau, Mchingiri, Mar. 1952, *Semsei* S688! & Lake Lutamba, 4 Sep. 1934, *Schlieben* 5243!
DISTR. **K** 7; **T** 6, 8; Mozambique
HAB. Coastal evergreen forest; 20–400 m

LOCAL USES. None recorded.
CONSERVATION NOTES. Apparently restricted to the increasingly scanty scraps of undisturbed coastal rainforest of East Africa. In Kenya known only from four collections, all in Kwale District; in Tanzania known only from four collections, all but one in Lindi District. In Mozambique, known from a single collection. This species is here assessed as VU B2 a, b(iii), or vulnerable, due to being known from less than 10 locations, with an area of occupancy of less than 2,000 km² and an inferred decline in its lowland moist forest habitat. *Robertson & Luke* 6315 (Gongoni Forest reserve): "Tree to 10 m which had been cut down for firewood by contractor (A Khan) to sell to Kenya Calcium Products, Waa". www.redlist.org already lists *S. schliebenii* as vulnerable (VU D2, based on an assessment by Lovett & Clarke in 1996).
NOTES. The leaves of *S. schliebenii* closely resemble those of *S. tragacantha*, a widespread species in Guineo-Congolian Africa. Confusion sometimes occurs. *S. tragacantha* differs in flowering with the leaves and in having the tips of the perianth united at anthesis, it also has a rather more tomentose lower surface to the leaf-blade. Luke (pers. comm.) notes that *S. schliebenii* grows above 400 m alt. in Mozambique.

5. **Sterculia appendiculata** *K.Schum.* in P.O.A. C: 272, t. 24 (1895) & in E.M. 5: 105, t. 9/c (1900); T.T.C.L.: 602 (1949); K.T.S.: 551 (1961); Wild in F.Z. 1: 554, t.104/B (1961); K.T.S.L.: 167, map (1994). Type: Tanzania, Lushoto District: Usambaras, *Holst* 2529 (K!, iso.)

Deciduous tree 12–40 m tall; bole straight, bark smooth, pale grey or greenish yellow or white, usually with short buttresses up to 20 m high, slash white; ultimate branchlets strongly longitudinally ridged, (3–)5–8(–18) mm thick, pale brown or grey brown, glabrous except the rusty, thickly woolly apex; bud-scales triangular, ± 3 mm long, 2 mm wide, rusty and scurfy when young. Leaf-blade ± orbicular in outline, shallowly (3–5–)7-lobed, 10–24 cm long, 7–19 cm wide, lobes widely based, acuminate, apical lobe 4–9 cm long, 5.5–11.5 cm wide, lateral lobes much smaller, 0.5–3.7 cm long, base deeply cordate, sinus 1.5–6 cm deep, edges never overlapping, glabrous above and below when adult, young leaves scurfy with dense rusty tomentum; petiole terete, sometimes slightly swollen at base and tip, 5.5–15.5 cm long, (0.7–)1–2 mm wide, glabrous when mature, scurfily brown-tomentose when young; stipules caducous. Inflorescence usually borne with the leaves, 1–14 per stem, each 4–11 cm long, 3–5 cm wide, thickly covered in yellow-brown stellate hairs; bearing 6–9 short branches, the lowest 2–22 mm from the base, 4–28 mm long, bearing 1–3(–5) flowers; pedicels 2(–12) mm long; 3 bracts at apex of pedicel, whorled, equal, subvalvate, calyx-like, elliptic, 7 mm long, 3 mm wide, velutinous, outer surface brown, inner purple grey. Flowers yellowish, perianth widely campanulate, 12–14 mm long, 18–23 mm wide, divided into 5 elliptic lobes, 6–8 mm long, 4–6.5 mm wide, indumentum as inflorescence, rusty yellow-brown, with stellate hairs, texture felty. Fruit with undehisced follicles ellipsoid, (5–)7–9 cm long, 3.5–6 cm wide, suture thickened, rostrum short and blunt, 3–5 mm long, stipe absent or very short and stout, 0–7 mm long, surface subscabrid with rusty brown dense stellate tomentum, pericarp woody, 2.5 mm thick increasing to 10 mm thick at the suture; follicle only opening slightly at dehiscence, inner surface glabrous, white, perhaps pulpy at maturity; seeds large, pale yellow, rounded oblong, 15–21 mm long, 10–13 mm wide, hilum ± rounded, 4–5 mm wide, white, lacking an aril, dangling from the fruit on a white thread 1–2 cm long. Fig. 1/2, 8, p. 6.

KENYA. Tana River District: 7 km NE Garsen, July 1972, *Gillett & Kibuwa* 19948! & Pokoma, Bura, 4 Jan. 1943, *Bally* 2036!; Kwale District: Mteza & Kwale, Aug. 1937, *Dale* 3772!
TANZANIA. Handeni District: Tamota, Jul. 1950, *Semsei* 593!; Iringa District: 5 km S confluence of Great Ruaha & Yovi Rivers, 7 Sep. 1970, *Thulin & Mhoro* 906!; Masasi District: 38 km Masasi–Tunduru, 19 Nov. 1967, *Gillett* 17934!; Zanzibar: near Pwani Mchangani, 1 Dec. 1930, *Greenway* 2613!
DISTR. **K** 7; **T** 3, 6–8, Z; Somalia, Malawi, Mozambique
HAB. Coastal and lowland riverine forest, often left standing on cultivated land, presumably for religious or superstitious reasons; 5–750 m

SYN. *Sterculia lindensis* Engl. in E.J. 39: 592 (1907); T.T.C.L.: 603 (1949). Type: Tanzania, Lindi District: Seliman–Mamba, *Busse* 2679 (B†, holo.; BR, EA (not found), iso.)

LOCAL USES. "Timber has been sawn for box staves at Mikesse, soft, perishable and requiring precautions against borers. Possible use for plywood?" *Wigg* 1350; "soft timber tree, wood easy to work, not very durable" *Semsei* 936; "leaves used as a side dish" *Barker* 464.
CONSERVATION NOTES. This species is here assessed as "Least Concern" in view of its large geographic range and because of its fairly common habitat.
NOTES. The palmately lobed leaves of *S. appendiculata* are similar in shape and in their large size to those of *S. quinqueloba*. The latter however, grows in drier forest and has a variegated trunk, the bark flaking off in plates, unlike the smooth whitish trunk of *S. appendiculata*. The leaves of S. *quinqueloba* are tomentose beneath, in contrast to the glabrous S. *appendiculata*. The similarity of the two species is only superficial, they differ very greatly in flowers and fruit.

Dorr has shown that *Sterculia lindensis* Engl., the placement of which has been much debated (e.g. Brenan 1956 loc. cit. placed it tentatively as a synonym of *Cola scheffleri* K. Schum.), is probably this species (Dorr in K.B. 59: 161 (2004)).

The lower altitudinal record was reported by Luke (pers. comm.), his record being "Luke sr Buxton Plot, Vipingo, Kilifi Distr.".

6. **Sterculia quinqueloba** (*Garcke*) *K.Schum.* in E.J. 15: 135 (1892) & in P.O.A. C: 271 (1895) & in E.M. 5: 104 (1900); T.T.C.L.: 602 (1949); Germain in F.C.B. 10: 267 (1963); Wild in F.Z. 1: 555, t. 104/E (1961); Troupin, Fl. Pl. lign. Rwanda: 667, t. 230/2 (1982). Types: Mozambique, Sena, *Peters* s.n. (B†, syn.); Macanga, *Peters* s.n. (B†, syn.)

Deciduous tree 2.7–21(–40) m tall, rarely a shrub, producing a clear hard gum; bole to 1 m in diameter, bark smooth and grey, often powdered with white dust and flaking in large plates, so variegated with white and red; slash orange or pink, sapwood white, exudate slight, watery; ultimate branchlets 9–18 mm thick, purplish grey; bud-scales triangular, 7–12 mm long, 3.5–7 mm wide. Leaf-blade orbicular in outline, strongly 5-lobed, 8–36 cm long, 10–34 cm wide, the lobes triangular–acuminate, subequal, the apical lobe 4.5–13 cm long, 3.5–10.5 cm wide, base deeply cordate, edges of the sinus overlapping, sinus 2–5 cm long, glabrous to subscabrid with sparse stellate and simple hairs above (densely tomentose when young), softly tomentose with sparse to very dense, fine stellate hairs beneath; petiole terete, 6–18 mm long, 2–4 mm wide, tomentose to pilose with a mixture of fine, small stellate and simple hairs with large, stout, pointed hairs filled with yellowish gum; stipules caducous. Inflorescence usually borne with the leaves, 3–8 per stem, each 15–32 cm long, 6–11 cm wide, indumentum sticky, as petiole, the fluid often exuded, the hairs appearing capitate; peduncle with numerous branches, 1.5–5(–7) mm thick at base, lowest branch 3–11 cm from the base, 7–12 cm long, with 7–9 partial peduncles, each (3–)5(–7)-flowered; bracts caducous, elliptic and acuminate, 3–7 mm long, 1–2.5 mm wide, velutinous; pedicels 1.5–5 mm long. Flowers with perianth pale or yellowish green, sweetly scented, campanulate, 3–4 mm long, 1.5–4 mm wide, (4–)5 triangular teeth each 1 mm long and wide, sometimes slightly reflexed, indumentum as inflorescence outside, inside largely of stout simple hairs. Male flowers with androphore ± 1.5 mm long. Fruit often with all 5 follicles developed, each cylindrical, 4.5–8.2 cm long, 1–1.3 cm wide, with a rostrum 0.3–1.6 cm long and a basal stipe 1.7–2.5 cm long, thickly tomentose, yellow-brown, sticky and fragrant; when dehisced 2.5–3 cm wide, revealing a mauve, tomentose inner surface; seeds ellipsoid, 7 mm long, 4–7 mm wide. Fig. 1/5, 11, p. 6.

TANZANIA. Shinyanga District: Shinyanga, *Koritschoner* 1984!; Tabora District: Tabora, Simbo Forest Reserve, 17 May 1977, *Ruffo* 943!; Songea District: Litenga Hill, 19 Apr. 1956, *Milne-Redhead & Taylor* 9779!
DISTR. **T** 1, 3–8; Congo-Kinshasa, Burundi, Angola, Zambia, Malawi, Mozambique, Zimbabwe
HAB. Dry, deciduous forest, woodland, wooded grassland or thicket, often on rocky slopes with *Brachystegia* or *Isoberlinia*; 90–1650 m

SYN. *Cola quinqeloba* Garcke in Peters, Reise Moss. Bot. 1: 130 (1861); Mast. in F.T.A. 1: 224 (1868)
   Sterculia livingstoneana Engl. in E.J. 39: 592 (1907) fide Wild in F.Z. 1: 555 (1961). Type: Zimbabwe, *Engler* 2936 (B, holo., probably destroyed)
   S. leguminosacea K.Schum. & Engl. in E.J. 39: 593 (1907); T.T.C.L.: 603 (1949). Types: Tanzania: Tabora, Meigwa, *Holtz* 1405 & Kilwa District: Mandandu, *Busse* 113 (both B, syn., probably destroyed), **syn. nov.**
   S. quinqueloba Sim, For. Fl. Port. E. Afr.: 18, t.6 (1909). Type: Mozambique, Maganja da Costa, *Sim* 998 (PRE, holo.)

LOCAL USES. Larger trees have been described as providing very hard timber for planks and sleepers (*Hendry* 528). However, this may well be a case of mistaken identity as other sources e.g. *Semsei* 848 describe the wood as soft and easy to work. "Wood soft, wood for making bee-hives" (*JC Newman* 140). The boiled bark is reported as being used as an enema for constipation (*Pirozynski* 360). "Food plant of chimpanzees" (*Uehara* 131); "leaves eaten by chimpanzees" (*Nishida* 82).
CONSERVATION NOTES. This species is here assessed as "Least Concern" in view of its large geographic range and because of its wide habitat range.
NOTES. Although *S. leguminosacea* is now known only from the original description drawn up from two collections which both lacked flowers and leaves, the fact that the follicles were described as glandular-tomentose and stipitate leaves no doubt that the true identity of this material is *S. quinqueloba*.
   S. quinqueloba is sometimes confused with the unrelated *S. appendiculata*. The differences are discussed under the latter.
   The lowest altitudinal record is derived from Luke (pers. comm.), who cites *Luke et. al.* 5588, Selous Game Reserve.

7. **Sterculia setigera** *Del.* in Voy. Méroé Bot.: 61 (1826); I.T.U.: 423 (1952); Germain in F.C.B. 10: 266 (1963); Vollesen in Fl. Eth. 2, 2: 183 (1995). Type: Sudan, Gonso (or Tertu), *Cailliaud* s.n. (MPU, holo.)

Large shrub or tree 4.5–12 m tall; bole often buttressed at the base; bark grey-purple, rough, flaking to leave pale patches; slash meat-red with paler streaks, watery, with white gum; ultimate branchlets yellowish brown tomentose to tomentellous, 3–7 mm wide, at length becoming ridged, rough, grey and exfoliating; bud-scales triangular, 2–2.5 mm long, 1.5–1.8 mm wide, tomentellous. Leaf-blade ± orbicular in outline, 3–5-lobed, 8.5–17 cm long, 8.5–17.5 cm wide, lobes subequal, wide, 4–7 cm long, 7–9 cm wide, apices acuminate, ± 3 cm long, 0.2 cm wide, base cordate, sinus 3–5 cm deep, edges overlapping, tomentellous above, densely yellow-brown tomentose with 6–8-armed pale stellate hairs below; petiole terete, 8–12 cm long, 2–3 mm wide, densely yellow-brown tomentellous; stipules caducous. Inflorescences borne 5–8 at the apex of leafless stems, 4–8.5 cm long, 2–4 cm wide, yellow-brown, tomentellous and densely stellate-hairy, with 3–10 lateral branches, lowest branch 3–9 mm from the base, 15–35 mm long, bearing 1–4 partial peduncles, each 1–3-flowered. Flowers with perianth green outside, red inside, widely campanulate, 9.5–13 mm long, 11–16 mm wide, divided into 5 oblong acute lobes, 5–9 mm long, 3–4 mm wide, outside velutinous, densely covered with short yellow-brown (when dried) stellate hairs which completely obscure the epidermis, inside glabrous apart from the upper $^2/_3$–$^3/_4$ of the lobes which more sparsely covered in patent white, simple hairs. Fruit with follicles ± ellipsoid, 7–12 cm long, 2–3 cm wide, nearly dehiscing flat, then slightly shorter, 4.5–8 cm wide, rostrum long, stout, straight or curved, 1–3 cm long, stipe short and stout, 0.2–0.5 cm long, pericarp woody, 2–4 mm thick, outer surface brown, tomentellous, inner surface yellow, drying dull orange, softly tomentose except the placenta which is covered in urticating yellow hairs ± 2 mm long; seeds with stout, peg-like stalks 2–3 mm long, persisting on old fruits; seeds ellipsoid, 11–12 mm long, 7–8 mm wide, grey-black, hilum sub-apical, round, 1 mm wide, at the margin of the aril; aril large, rounded, apical, 5 mm wide, 2 mm high, red, drying orange or red. Fig. 2, p. 15.

UGANDA. Acholi District: Gulu–Pakelli, *Eggeling* 1226!; Karamoja District: Labwor, 4 Jun. 1940, *Thomas* 3709!; Bunyoro District: Bunyoro, near Lake Albert, *Dawe* 771!
TANZANIA. Shinyanga District: Shinyanga, Mar. 1936, *Burtt* 5608!
DISTR. **U** 1, 2; **T** 1; Senegal, Gambia, Ghana, Togo, Benin, Nigeria, Cameroon, Central African Republic, Congo-Kinshasa, Sudan, Ethiopia, Angola
HAB. Wooded grassland; 750–1200 m

SYN. *Sterculia tomentosa* Guill. & Perr. in Fl. Senegambia Tent. 1: 81, t. 16 (1831); Mast. in F.T.A. 1: 217 (1868); K. Schum. in E.M. 5: 106 (1900), *non* Thunberg (1802). Type: Senegal, Walo, Dagagna; Galam, Bakal, *Leprieur* s.n. (P, holo.)
  Cola tomentosa (Guill. & Perr.) Schott. & Endl., Meletem.: 33 (1832)
  Clompanus tomentosa (Guill. & Perr.) Kuntze, Rev. Gen.: 78 (1891)

LOCAL USES. Wood white and soft, of little value. Seeds taste like groundnuts, eaten in Uganda (fide Eggeling in I.T.U.: 423 (1952)).
CONSERVATION NOTES. This species is here assessed as "Least Concern" in view of its large geographic range and because of its common habitat.
NOTES. *Sterculia setigera* is closely related to *Sterculia africana*, the flowers being very similar apart from indumentum, the fruits being almost identical. It is distinguished from the latter by the 6–8-armed stellate hairs that make the lower leaf-blade tomentose and the perianth yellow-brown velvety, and are so dense as to entirely obscure the epidermis below. In addition, the leaves are at least twice the size usual in *Sterculia africana*. However, the latter is a variable species and some larger leaved specimens from **T** 3 with thicker indumentum than the norm have been treated, for example by Brenan (T.T.C.L.: 603 (1949)) as *S. setigera*.
  Throughout its range, *S. setigera* seems to be rather constant in leaf shape and size, and in perianth indumentum. A discordant element, with the indumentum of the lower leaf-blade silkily white, occurs in Ethiopia. This seems referable to *S. cinerea* A.Rich., normally treated as a synonym of *S. setigera*. Further investigation is warranted.

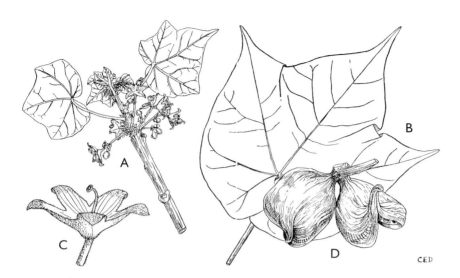

FIG. 2. *STERCULIA SETIGERA* — **A**, habit × ³/₅; **B**, leaf × ³/₅; **C**, male flower × 3; **D**, fruit × ²/₃. Reproduced from Nigerian Trees 1: t. 59 (1960). Drawn by Christine Darter.

8. **Sterculia mhosya** *Engl.,* V.E. 3 (2): 455, t. 208/A–H (1921); T.T.C.L.: 602 (1949); Wild in F.Z. 1: 555, t. 104/F (1961). Types: Tanzania, Kahama District: Usumbwa [Ussumbwa] and Dodoma/Mpwapwa District: Ugogo, Tabora, collector not indicated but probably *Braun* (presumably B†, holo.); Lushoto District: Usambara Nyembe-Bubungwa, *Hammirstein* 3079 (EA!, lecto., selected here)

Tree, rarely a shrub, 0.9–10 m tall; bole smooth, bark sometimes slightly corrugated, greyish or greenish, papery, thinly flaking; ultimate branchlets usually glossy coppery red, 2.5–6 mm thick, mostly glabrous, but tomentose towards the tip. Leaf-blade ± orbicular in outline, 3–5(–7)-lobed for half its length, rarely entire, (5–)7–12(–19) cm long, (4.5–)7–16(–22) cm wide, ovate-acuminate, the lobes subequal or the basal lobes partly to completely reduced, usually rounded, 2–7.5 cm long, 2–8 cm long, with a narrowly acuminate tip 1–2.5(–4) cm long, 0.2–0.6(–1) cm wide, base deeply cordate, sinus (0.8–)1–3(–4.5) cm deep, the edges rarely overlapping, margin sometimes repand to rounded serrate, texture papery to thinly leathery, glabrous to subscabrid above, puberulent to subscabrid, rarely glabrous, with scattered stellate hairs beneath; petiole terete, (2–)3–8(–13.5) mm long, (0.5–)1(–2) mm wide, tomentellous, often with simple, purple glandular hairs amongst the more numerous stellate hairs, or glabrous; stipules caducous. Inflorescences borne from the stem apex (or short spur-shoots near the stem apex) when leafless or nearly so, 1–6(–10) per stem, sticky, covered with large purple glandular hairs mixed with white stellate hairs, each inflorescence (3–)5–10(–14) cm long, 1.5–3.5 cm wide, bearing 3–9 branches, each 1(–3)-flowered; pedicels 3–7 mm long. Flowers with perianth yellow-green to dull red or crimson, widely campanulate, (7–)9–12(–14) mm long, (8–)11–16(–19) mm wide, divided into 5 triangular-acute lobes 5–9.5 mm long, 2–4 mm wide, outer surface with indumentum as the inflorescence. Fruits with follicles cylindrical, 6–9 cm long, 1–2.4 cm wide, dehiscing flat, then slightly shorter, 2.2–3.5 cm wide, rostrum stout, (0.3–)1 cm long, stipe absent or very short, 0.1 cm long, pericarp rather thin and leathery, 1 mm thick, outer surface dull pink, drying brown, tomentose to tomentellous; seeds ellipsoid, 9–11 mm long, 5–6.5 mm long, grey-black, hilum sub-apical, round, 1 mm wide, at the margin and partly occluded by the apical aril 2–3.5 mm wide, 1.5–2 mm high, orange, drying orange or white. Fig. 1/6, 12, p. 6.

Tanzania. Mwanza District: Mwanza–Airport, Ilemela village, 18 Feb. 1982, *Mwasumbi* in MZ 306!; Tabora District: 20 km Tabora–Ugalla, 26 June 1949, *Hoyle* 1044!; Kondoa District: Mondo, 17 May 1978, *Ruffo* 1315!

Distr. **T** 1, 3–5, 7; Zambia

Hab. Dry bushland or woodland with *Acacia*, *Commiphora* and *Combretum*, often on rocky hillsides; 700–1300 m

Syn. *S. burttii* Bruce in Hora, T.T.C.L.: 173 (1940) fide Brenan in T.T.C.L.: 604 (1949), nomen, **syn. nov.**
    Sterculia sp. nr. *mhosya* Brenan, T.T.C.L.: 602 (1949). Based on: Tanzania, Mpwapwa, *Burtt* 2412, 2513, 3927!
    S. sp. Brenan, T.T.C.L.: 604 (1949). Based on: Tanzania, Singida, *Burtt* 745!

Local uses. Bark commonly used as amulet string, on the wrist in children against disease (*Gane* 24), also used for making rope (*Ruffo* 1483) and as a decoction against indigestion. Leaves used as toilet paper for children; boles for canoes but plants of enough size hard to obtain (*Azuma* 563). Seeds edible, used as groundnuts (*Ruffo* 1315, 1483 and others).

Conservation notes. This species is here assessed as "Least Concern" in view of its large geographic range and because of its wide habitat range.

Notes. Most of the abundant material of *S. mhosya* from Tanzania has until now borne the name *S.* sp. nr. *mhosya*, bestowed by Brenan (T.T.C.L.: 602 (1949)). Brenan pointed out that although Engler's plate of *S. mhosya* matched the material available to him, Engler's description (at odds with the plate) mentioned a stipe 1 cm long (not seen in such East African material). He concluded that two species were involved, but pointed out that if the description of it as stipitate was in error, as now seems likely (no such stipitate specimens have come to light), then 'what I have called *S.* sp. nr. *mhosya* and *S. mhosya* are presumably the same'. I have no doubt that this is the case. It is quite concievable that Engler mistakenly described as a basal stipe the terminal rostrum of the fruitlet, which is usually 1 cm long in this species.

    The protologue of *S. mhosya* Engl. does not cite specimens and in any case material at B is believed destroyed so a lectotype is necessary. Two specimens present themselves as candidates: *Braun* 5390 (collected 1913) and *Hammirstein* 3079 (1910), both being made for the Amani herbarium (specimens since transferred to EA and K) and so likely to have been available to Engler in drawing up his protologue. Moreover both have localities mentioned in the protologue. The second specimen is selected because it bears the unusual orthography of the local name used by Engler, "mhosya".

    Brenan's (T.T.C.L.: 604 (1949)) thirteenth *Sterculia*, '*S.* sp.' (*S. burtti*) is based merely on an aberrant, poorly lobed specimen of *S. mhosya*. The fruits, and the presence of purple glandular hairs which so clearly characterize this species indicate its true identity.

9. **Sterculia stenocarpa** Winkler in F.R. 18: 123 (1922); T.T.C.L.: 604 (1949); K.T.S.: 552 (1961); K.T.S.L.: 168, map (1994); Vollesen in Fl. Eth. 2, 2: 184, fig. 80.8: 10 (1995); Thulin, Fl. Somal. 2: 25 (1999). Type: Kenya, Teita District: between Taveta River and Voi, *Winkler* 4047 (WRSL, holo.)

Tree, rarely a shrub, 4–12 m tall, spreading, the crown often as wide as the tree is high, usually with numerous branches from shortly above ground; bole often short and stout, bark smooth, grey or reddish brown; ultimate branchlets orange-brown, 1.5–3.5 mm wide, densely hairy long-tomentose with pale yellow stellate hairs. Leaf-blade ± orbicular in outline, usually shallowly 3-lobed, divided for ± ¹⁄₄ sometimes 5–7-lobed, rarely entire and then usually with a repand margin, (2.5–)4–7(–12) cm long, (2–)4–10(–13) cm wide, the lobes subequal, usually wider than long, 1.4–3.4 cm long, 2.2–6 cm wide, apex rounded, sometimes shortly subacuminate, base deeply cordate, the sinus (0.5–)0.7–1.4(–3) cm deep, the edges not quite overlapping, long-tomentose above and beneath with indumentum as the stem; petiole orange brown, terete, (1–)2–6.5(–8) cm long, 1–1.5 mm wide, indumentum as the stem; stipules persistent, linear-triangular, 5–12(–15) mm long, 1–1.5(–2) mm wide at the base, black, outer surface often glabrous, inner surface densely long-tomentose with white hairs. Inflorescences usually borne on leafless stems, in large numbers, ± 15–25 from the apex of the main shoot, or 1–4 from the apex of the spur shoots, 0.7–1.7 cm long; pedicels 3–6 mm long. Flowers with perianth yellow green, striped inside with red, sometimes altogether dark red, widely campanulate, tube 7–12 mm long, (9–)12–18(–28) mm wide, divided into 5 spreading triangular teeth, each 8–14 mm long, 4–7 mm wide, outside densely long-tomentose, indumentum as the stem. Fruits with follicles pendant, cylindrical, moniliform, 6.5–9 cm long, 0.8–1.2 cm wide, dehiscing completely flat, then slightly shorter, 4.5–7.2 cm long, 1.8–2.6 cm wide, rostrum curved, 0.2–1.7 cm long, stipe absent, pericarp leathery, ± 0.5 mm thick, occasionally slightly woody and ± 1 mm thick, outer surface usually grey-green, sometimes reddish, tomentellous to subscabrid; seeds with stalks small and inconspicuous, glabrous, 0.5–1 mm long and wide; seeds ellipsoid, 8 mm long, 5 mm wide, grey-black, aril apical, rounded, 2–4 mm wide, 1–2.5 mm high, largely obscuring the round hilum 0.5–1 mm wide

UGANDA. Karamoja District: Rupa, Aug. 1955, *J. Wilson* 167!
KENYA. Northern Frontier District: Dandu, 5 June 1952, *Gillett* 13153!; Machakos District: 18 km
    Mtito Andei–Mombasa, 9 Oct. 1962, *Greenway* 10811!; Voi District: Tsavo National Park E, Voi
    Gate Camp Site, 7 Dec. 1966, *Greenway & Kanuri* 12676!
TANZANIA. Mbulu District: above Msasa Park Wardens Camp, 24 Mar. 1964, *Greenway & Kanuri*
    11407!; Pare District: Kisengero–Lembene, 31 Jan. 1936, *Greenway* 4545! & Same, 18 Sep.
    1987, *Ruffo* 2559!
DISTR. U 1; K 1–4, 6, 7; T 2, 3, 7; Sudan, Ethiopia, Somalia
HAB. Dry bushland with *Commiphora, Terminalia, Combretum* and *Acacia*; 0–1700 m

LOCAL USES. Seeds edible, *J. Wilson* 167; bark used for string, tough and durable; *Graham* 1712;
    "used by Somalis for string, rope etc." *Sampson* 81.
CONSERVATION NOTES. This species is here assessed as "Least Concern" in view of its large
    geographic range and because of its wide habitat range.
NOTES. It is noticeable that the few specimens from T 2, Tanzania (10–12 m tall) are all
    reported as being taller than other specimens from East Africa (usually 6–8 m tall). Fruits
    from NW Kenya generally have pericarps about twice as thick as the more coastal
    populations.
    This species is easily recognized when sterile on account of the persistent stipules, not
otherwise known in East African *Sterculia*. These are particularly prominent on the short spur-
shoots which also bear the remains of the inflorescences of previous seasons. This character,
and the long-tomentose, patent-stellate indumentum of stem, leaf-blade, petiole,
inflorescence and perianth readily serve to distinguish specimens from the sympatric
Sterculia africana and *S. rhynchocarpa* with which it has often been confused. The narrow,
cylindrical, moniliform fruit with a leathery pericarp which dehisces completely flat is very
different to those of the latter and suggests that its closest affinity lies with *S. mhosya*
(Tanzania), which is readily distinguished by the acuminate leaf-blade lobes and
inflorescence with purple glandular hairs.
    The lowest altitudinal record is derived from Luke (pers. comm.), who cites his sight
record at Shimoni Village, Kwale Distr.

10. **Sterculia rhynchocarpa** *K.Schum.* in E.J. 34: 323 (1904); T.T.C.L.: 603 (1949); I.T.U.: 423 (1951); E.P.A., 1: 585 (1958); K.T.S.: 551 (1963); Blundell, Wild Fl. E Africa: 73, t. 444 (1987); Vollesen in Fl. Eth. 2, 2: 184, fig. 80.8: 9 (1995). Type: Tanzania, Pare District: Pare Mts, Gonja–Kisiwani [Kisuani], *Engler* 1523, 1563 (both B, syn., probably destroyed)

Shrub or small tree, 0.6–9(–15) m tall, often spreading from or near the base; bole often swollen, bark smooth, purplish, grey or reddish brown, sometimes peeling; young extension shoots rare, pale brown, ± 2.5 mm wide, shortly tomentellous, older stems and spur shoots very pale grey, smooth and finely ridged, 3–6 mm wide; bud-scales triangular, 1–3 mm long, 0.5–1.5 mm wide, inside densely white tomentose. Leaf-blade orbicular to ovate in outline, shallowly 3-lobed or entire, (0.5–)1–5(–9) cm long, (0.5–)0.8–5(–8.5) cm wide, lateral lobes always less deep than the apical lobe, apex rounded to subacuminate, base truncate to cordate, edges never meeting, subtomentellous to glabrescent above, softly tomentose to tomentellous, rarely glabrescent beneath; petiole terete, orange-brown, (0.4–)0.7–2(–3.7) cm long, subtomentellous with minute stellate hairs or glabrescent. Stipules caducous. Inflorescences borne on ± leafless stem apices, 1–2(–10) per shoot, 1.5–3.5(–4.8) cm long, 1–2 cm wide, indumentum subtomentellous, branches 3–6(–10); pedicels 1.5–6(–10) mm long. Flowers green, with red flush and stripes inside, widely campanulate, (5–)6–11(–13) mm long, (8–)10–14 mm wide, divided into 5 triangular, reflexed lobes (4–)6–8 mm long, (3–)4–5 mm wide, outside with scattered minute stellate hairs, inside glabrous apart from the long-hairy lobes. Fruits with follicles ± ellipsoid in lateral view, 4–11 cm long, widest at the suture side in end view, 2–3.5 cm wide, dehiscing by 90–180 degrees, then slightly shorter and 4–6 cm wide, rostrum stout, often very long, and curved at the tip, (1–)2–4.5 cm long, stipe very short or absent, pericarp woody, 1.5–3 mm thick, outer surface grey-green, subtomentellous, with sharply raised longitudinal ridges or distinctly prickly, prickles 1–3 mm high; seeds with peg-like stalks 2–4 mm long, 1 mm wide, covered with urticating hairs, persistently attached to the placenta; seeds ellipsoid–oblong, grey-black, 12–14 mm long, 5–7 mm wide, the aril apical, rounded, orange yellow or red, 2–4 mm wide, 1–2 mm high; hilum at the aril margin, round, ± 1 mm wide. Fig. 1/4, 10, p. 6.

KENYA. Northern Frontier District: Ijara, Marodi, 7 Jan. 1943, *Bally* 2059!; Kitui District: Nuu, SE Nyaani, *Kuchar* 14996!; Tana River District: Nairobi–Garissa, 25 km E Hatama Corner, 9 May 1974, *Gillett & Gachathi* 20541!
TANZANIA. Lushoto District: 8 km SE Mkomazi, 2 May 1953, *Drummond & Hemsley* 2378!; Pangani District: Msubugwe Forest Reserve, 14 Mar. 1963, *Mgaza* 570!; Tanga District: Kirindami, 3 Feb. 1966, *Faulkner* 3740!
DISTR. **K** 1, 4, 7; **T** 3; Ethiopia, Somalia
HAB. Dry bushland or woodland, usually with *Acacia* and *Commiphora*; 5–1000 m

SYN. *S. rivae* Chiov. in Result Sci. Miss. Stef.-Paoli, Coll. Bot. 1: 203 (1916). Type: Ethiopia, Gamo Gofo/Sidamo Sagan River, *Ruspoli & Riva* 1767 (1695) (FT, holo.)
    S. africana sensu K.T.S.L.: 167 (1994)

LOCAL USES. Bark used for binding fibre, *Boy Joana* 7430; seeds eaten, *Joy Adamson* 288.
CONSERVATION NOTES. This species is here assessed as "Least Concern" in view of its large geographic range and still common habitat.
NOTES. *Sterculia rhynchocarpa* is very closely related to *Sterculia africana*. The leaves of the latter are generally larger than those of the former, but there is an area of overlap at which point it is very difficult to identify a specimen to species without there being fruits present, the flowers of the two species being ± identical. For this reason it is especially important to collect fruits when making collections. Since the pericarp is so woody, it is likely that if fruits are not present on the tree, they will be found on the ground throughout the year. *S. rhynchocarpa* is the only member of the genus in tropical Africa with warty or spiny fruits, though two more such species occur in S Africa. *S. rhynchocarpa* can also be distinguished from *Sterculia africana* by the grey-green, subtomentellous fruits (versus yellowish brown tomentose) generally with a much longer rostrum.

11. **Sterculia africana** (*Lour.*) *Fiori* in Agric. Colon. Ital. 5, suppl.: 37 (1912); T.T.C.L.: 602 (1949); E.P.A. 1: 584 (1958); Wild in F.Z. 1: 553 (1961); K.T.S.: 551, t. 101 (1961); K.T.S.L.: 167, fig., map (1994); Vollesen in Fl. Eth. 2, 2: 184, fig. 80.8: 1–7 (1995); Thulin, Fl. Somal. 2: 35 (1999). Type: Mozambique, Mossuril, *Loureiro* s.n. (P!, holo.)

Tree 4–10(–18) m tall; bole often thick and squat, bark whitish grey or liver-colored; slash unknown; young extension shoots pale brown, shortly tomentellous. Leaf-blade orbicular to ovate in outline, shallowly 3-lobed or entire, 3.5–10(–12.5) cm long, (3–)5–8(–11) cm wide, lateral lobes 0.5(–2) cm deep, always more shallow than the apical lobe, apex rounded to acuminate, base cordate, sinus 0.5–1.5 cm deep, edges usually not quite meeting, shortly and thinly tomentose to glabrescent above and beneath, hairs stellate, greyish with 5–7 ± horizontal arms; petiole 2.3–7.5(–12.5) cm long, 0.5 mm thick, tomentellous with greyish stellate hairs; stipules not long persistent. Inflorescences borne on ± leafless stem apices, 2–6(–20) per shoot, 1.5–3 cm long, 1–2 cm wide, indumentum as the leaf, spike-like or branches 2–5; pedicels 3–8 mm long. Flowers greenish or yellowish with pink or red markings, widely campanulate, 7–8 mm long, (5.5–)8–20 mm wide, divided into 5 rounded-triangular, patent or reflexed lobes 7–9 mm long, 4–4.5 mm wide, outside with small stellate hairs, as the leaf, inside glabrous apart from the lobes which densely covered in longer, white silky simple to 3–4-armed stellate hairs. Fruits with follicles ± ellipsoid in lateral view, 6–11 cm long, widest at the equator in end view, 5–7 cm wide, dehiscing by 90–180°, then shorter and 8–9 cm wide, rostrum, if present, stout, slightly curved, 0–2 cm long, stipe stout and short if present, 0–0.5(–0.8) cm long, pericarp woody, 1.5–2(–4) mm thick, outer surface yellowish brown, tomentose to subscabrid, without sculpturing, inner surface yellowish white, softly and thinly tomentose, characteristically lined, placenta covered densely in yellow-brown urticating hairs 2–3 mm long; seeds with peg-like stalks (2–)3–3.5 mm long, (0.5–)1–1.5 mm wide, covered with urticating hairs, persistently attached to the placenta; seeds ellipsoid–oblong, grey-black, 12–15 mm long, 7–8 mm wide, the aril apical, 2–3 mm long, 3–5 mm wide, drying white; hilum at aril margin, round, 1–1.5 mm wide. Fig. 1/1, 7, p. 6.

TANZANIA. Handeni District: Zindeni Hill, 12 Sep. 1933, *Burtt* 4880!; Mpwapwa District: road to Mlunduzi, 12 May 1976, *Magogo & Ruffo* 686!; Morogoro District: Morogoro, Sep. 1951, *Eggeling* 6294!; Zanzibar, 1927, *Toms* 257!
DISTR. **T** 3, 5–8, **Z**; Sudan, Ethiopia, Somalia, Angola, Zambia, Malawi, Mozambique, Zimbabwe
HAB. Dry bushland or grassland with *Combretum* or *Acacia*, at the sea-shore in bushland with *Sideroxylon* and *Xylocarpus*, often on coral rock; 0–600 m

SYN. *Triphaca africana* Lour., Fl. Cochinch.: 577 (1790)
    Sterculia triphaca (Lour.) R.Br. in Bennett, Pl. Jav. Rar.: 228 (1844); Mast. in F.T.A. 1: 216 (1868); K. Schum. in E.M. 5: 106 (1900) pro parte; Engl. in V.E. 3 (2): 452 (1921). Type: as for *S. africana*
    S. ipomoeifolia Garcke in Peters, Reise Mossamb. Bot. 1: 130 (1861). Type: Mozambique, Sena, *Peters* (B, holo. probably destroyed)
    S. triphaca R.Br. var. *rivaei* K.Schum. in E.M. 5: 106 (1900). Types: Somalia, Savati, *Riva* 1597 ((B, syn., probably destroyed, FT); Kenya?, Muansa, fl. May 1892, *Stuhlman* 4587 (B, syn., probably destroyed); Teita District: Ndi (Taita), fl. Feb. 1877, *Hildebrandt* 2566 (B, syn., probably destroyed); Tanzania, Lushoto District: Usambara Mts, *Holst* 2373 (B, syn., probably destroyed; K!, isosyn.); Malawi, *Buchanan* 1025 (B, syn., probably destroyed); Angola, *Buchner* 518 (B, syn., probably destroyed; K!, isosyn.)
    S. setigera sensu Brenan in T.T.C.L.: 602 (1949), *non* Del.

LOCAL USES. Bark used for rope, *Greenway* 5110.
CONSERVATION NOTES. This species is here assessed as "Least Concern" in view of its large geographic range and because of its wide habitat range.
NOTES. It is remarkable that though *Sterculia africana* is the only member of that genus illustrated in 'Kenya Trees and Shrubs', this species is not known from that country, though it is recorded to the North, from Ethiopia and Somalia, and to the South from Tanzania to South Africa and Botswana. Many specimens from Kenya which have initially borne the name *Sterculia africana* have subsequently proven to be the closely related *S. rhynchocarpa* (for diagnostic characteristics, see there). Both *Greenway* 5110 & 5289, figured in the K.T.S. plate, are from Mafia Island, Tanzania.
    Sterculia triphaca var. *rivaei* was very probably based on specimens attributable to both *Sterculia africana* and to *S. rhynchocarpa*. This is because two of the six syntypes of var. *rivaei* (both believed destroyed) were from Kenya, whence specimens of *S. africana* are unknown, although *S. rhynchocarpa* is common.
    Sterculia arabica (R.Br.) T.Anders of Yemen and Oman is very closely related to *Sterculia africana*. Specimens of the former can usually be differentiated by the smaller, uniformly rounded leaves and smaller fruits, about half the size of the latter. However, qualitative differential characters seem scarce and monographic research might relegate these two taxa to subspecific rank. *Sterculia arabica* has nomenclatural priority.
    Specimens collected from coastal districts at sea-level, including all those from Mafia and Zanzibar, have very much larger and more sparsely hairy leaves than those from Acacia

bushland, but this phenomenon may be environmentally engendered. Although several varieties have been recognized throughout the coastal range of *Sterculia africana*, examination of the available material from E Africa suggests that these are unwarranted. Although the species does seem rather variable in leaf-shape and indumentum, intermediates can be demonstrated.

The record from **T** 8 is derived from a sight record at Ngarama North Forest Reserve by Luke (pers. comm.).

Sterculia foetida *L.*, Sp. Pl. 1008 (1753); K. Schum. in P.O.A. C: 271 (1895); U.O.P.Z.: 455 (1949); I.T.U: 65 (1953); Wild in F.Z. 1(2): 557 (1961); Wild & Gonç. in F.M. 27: 45 (1979)

Leaves digitately 5–7-foliolate. Flowers foetid red and 2.5 cm wide.

Originally from India but long cultivated in East Africa, largely in the botanic gardens of Entebbe (*Chandler* 1423), Nairobi, Amani and Dar-es-Salaam (*Chilongola* 120).

2. COLA

Schott & Endl., Melet. Bot. 33 (1832)

Monoecious perhaps sometimes dioecious, evergreen or rarely deciduous trees or shrubs. Stems often white and glabrous; bud-scales usually markedly caducous. Leaves simple, elliptic to oblanceolate, rarely ± orbicular and then digitately lobed, usually entirely glabrous, entire, venation finely reticulate; petiole usually swollen and kneed at base and apex, usually very variable in length on a single stem. Stipules usually caducous. Inflorescences often from leafless axils of the stem, sometimes cauliflorous, short racemose, paniculate or fasciculate with flowers each on an articulated stalk (the upper part the pedicel, the lower peduncle) with bracts basal and often deeply bilobed. Flowers with a single perianth whorl united at the base, valvate in bud. Female flowers with apparently indehiscent, subsessile stamens forming a ring around the base of the ovary; ovary subsessile, subglobose or ellipsoid, 3–5-locular, densely hairy, the style usually very short, the stigma with pronounced lobes, as many as the carpels. Male flowers with androphore and head of 4–12 anthers, each with 2 separate thecae in one, or two (*C. gigantea*) rows, gynoecium vestigial, inconspicuous. Fruit carpels separated, each developing into indehiscent follicles or berries, each ± globose to cylindrical, pericarp firm or fleshy, inner surface glabrous, bearing 1–10 seeds. Seeds lacking an aril, embryo orientated towards hilum.

About 125 species, restricted to evergreen forest in continental Africa. Most numerous in Guineo-Congolian Africa, but with an important secondary area of diversity in coastal eastern Africa (± 20 native species) where many of the taxa have very limited distribution, each sometimes occurring in only one or two forests. This account is based on the work of Brenan who described nearly half the species in East Africa and whose account of 'The genus *Cola* in Kenya, Uganda and Tanganyika' (K.B. 11: 141, 1956) was a major step in unravelling the species of this area. In the last twenty years, since Brenan ceased work on *Cola*, botanical inventory work in the lowland coastal forests of Tanzania, notably by Quentin Luke, has brought to light five new species, four of which are newly described in this account. In addition a new name is provided for the Kenyan species previously known as *C. clavata*.

Although in the allied genus *Sterculia*, vegetative characters, particularly leaf indumentum, are always very useful for distinguishing species, this is not so in *Cola* where vegetative characters are generally rather uniform. Intermittent flowering has further hampered studies in the genus. In East Africa trees often seem not to flower every year (Vollesen pers. comm.). Many species are known from very few fertile collections and in several species fruits are unknown. Some apparently dioecious trees can also produce flowers of the opposite sex earlier or later in the season or from different parts of the stem (see Brenan in K.B. 40: 87 (1985)). Many species of

Cola are separated by obscure diagnostic floral characters and are consequently frequently misidentified. The key below is based on vegetative characters as far as possible. In the absence of flowers, geographical information can often be used to arrive at a provisional species name (and for this reason is included in the key) since many of the species in East Africa are restricted in distribution. Otherwise, sterile collections are often impossible to identify even by the specialist.

1. Leaves in whorls of 3 or 4; cultivated . . . . .   *C. verticillata* (p. 24)
   Leaves alternate or occasionally subopposite 2
2. Deciduous tree; leaf undersurface white,
   thickly tomentose; **U** 1, 2, 4 . . . .   1. *C. gigantea* (p. 24)
   Evergreen; leaf undersurface drying green
   or brown, glabrous or (*C. pierlotii*) very
   sparsely stellate-hairy 3
3. Leaves digitately lobed (but sometimes
   mixed with entire leaves, especially on
   fertile shoots) 4
   Leaves never lobed 5
4. Leaves usually 3(–5)-lobed, and fertile shoots
   sometimes with some unlobed leaves; **T** 3, 6   3. *C. scheffleri* (p. 25)
   Leaves usually 5-lobed; cultivated; **K** 4 . . . . .   *C. millenii* (p. 23)
5. Petiole rather uniformly short, up to 18
   (–32) mm long 6
   Petiole very variable in length, on a single
   stem from 4–5 mm to at least 30 mm long,
   in most species to 60–100(–270) mm long 9
6. Petiole 1.5–18(–32) mm long; flowers 4–5 mm
   across; perianth lobes 2.2–3 mm long 7
   Petiole 2–7 mm long; flowers 6–9.5 mm across;
   perianth lobes 3–4 mm long 8
7. Leaf-blade 1.9–10(–12) cm long, base with
   straight edges; flower stalks 3–4 mm;
   perianth lobes hairy on inner surface;
   K 7, **T** 3 . . . .   15. *C. minor* (p. 38)
   Leaf-blade (2.5–)9–16 cm long, base with
   concave edges; flower stalks 8–9(–12) mm
   long; perianth lobes glabrous on inner
   surface; **T** 6 . . . .   16. *C. kimbozensis* (p. 39)
8. Flowers whitish, yellowish or pinkish brown;
   basal bracts 3–4, conspicuous, 1.7–2.5 ×
   2–3 mm; perianth tube cup-shaped, ± 2 mm
   diameter, perianth lobes patent; 30–500 m;
   K 7, **T** 3, ?6 . . . .   12. *C. uloloma* (p. 36)
   Flowers green, basal bracts inconspicuous;
   pedicels 0.2–0.3 mm diameter; perianth
   tube bowl-shaped, ± 3.5 mm diameter,
   perianth lobes ascending; 1500 m; **T** 4 & 7   13. *C. chlorantha* (p. 37)
9. Flowers borne on persistent woody bosses on
   the trunk and on old branches (not
   merely below the leaves); perianth of male
   flowers held flat, not cupped, completely
   deep pink to red, drying purple (rarely
   (*C. lukei*) cream or pale yellow 10
   Flowers borne amongst or just below the
   leaves; perianth of male flowers not flat,
   but ± cupped, white, brownish, yellow
   greenish, sometimes with red markings 14

10. Flowers cream or pale yellow; **T** 3 . . . .   6. *C. lukei* (p. 28)
    Flowers pink-purple, purple or red 11
11. Pedicels 3–4 mm long; lowland Tanzania
    (below 650 m) 12
    Pedicels (6–)10–40 mm long; Kenya and
    montane (above 850 m) Tanzania 13
12. Flowers per fascicle numerous; inner
    perianth densely stellate-hairy; androphore
    glabrous; **T** 8 (Rondo Plateau) . . . .   7. *C. rondoensis* (p. 30)
    Flowers 1–4 per fascicle; inner perianth
    glabrous; androphore stellate-hairy; **T** 6
    (Kimboza Forest) . . . .   8. *C. quentinii* (p. 31)
13. Pedicels (13–)17–40 mm long, pedicels and
    perianth glabrous; styles 3–4; **K** 7, **T** 3 . . .   5. *C. porphyrantha* (p. 27)
    Pedicels (6–)10–20(–28) mm long, pedicel
    and outer perianth usually with scattered
    stellate hairs (rarely dense, or glabrous);
    styles 5; **T** 6–7 . . . .   4. *C. stelechantha* (p. 26)
14. Bud-scales persistent (but caducous in *C.
    ruawaensis*); perianth lobes 6 or more,
    10–20 mm long; on limestone in **K** 7 and
    T 8 15
    Bud-scales falling very early; perianth lobes 5
    (rarely more), <10 mm long (but 10–14 mm
    in *C. congolana*); throughout E Africa, but
    not on limestone 16
15. Bud scales entire, acuminate; stem tomentum
    persistent; stipules persistent; **K** 7 . . . . . .   9. *C. octoloboides* (p. 33)
    Bud-scales notched; stems glabrous or with
    tomentum caducous; stipules caducous; **T** 8
    (Ruawa Forest) . . . .   10. *C. ruawaensis* (p. 34)
16. Flowers in short racemes 17
    Flowers single or in fascicles 20
17. Surface of apical bud, and outer surface of
    perianth, completely covered in soft, felty,
    golden brown hairs; some leaves usually
    lobed; **T** 3, 6, 7 . . . .   3. *C. scheffleri* (p. 25)
    Surface of apical bud, and outer surface of
    perianth glabrous, or white stellate-hairy;
    leaves always entire 18
18. Younger leaves sparsely white stellate-hairy
    on both sides; axillary buds spiny;
    perianth lobes very thick; **U** 2 . . . .   2. *C. pierlotii* (p. 24)
    Young leaves glabrous; axillary buds blunt,
    perianth lobes membranous; cultivated 19
19. Leaves long-acuminate, often slightly twisted;
    fruits not rugose or tuberculate, scabrid;
    cultivated . . . .   *C. acuminata* (p. 23)
    Leaves abruptly acuminate, not twisted; fruits
    rugose or tuberculate, glabrous; cultivated   *C. nitida* (p. 23)
20. Bracts to 7.5 mm long, 9.5 mm wide; perianth
    10–14 mm long; **U** 2 . . . .   11. *C. congolana* (p. 34)
    Bracts to 2.5(–4.5) mm long, 1.5 mm wide;
    perianth < 8 mm long 21

21. Young stems and petiole usually with orange
or reddish brown, scurfy hairs persisting;
pedicel and outer perianth thickly covered
in hairs ± 0.5 mm long of same colour;
T 6–8 22
Young stems and petiole usually white and
glabrous, sometimes with sparse, grey or
blackish, scabrid indumentum 23

22. Longest petiole to 6.7 cm long; flowers borne
mostly below the leaves; pedicels articulated
± $\frac{1}{3}$ of its length from the base; hairs long,
grey, shaggy, 0.4–0.6 mm long; androphore
glabrous .... 17. *C. microcarpa* (p. 40)
Longest petiole to 10 cm long; flowers mostly
among the leaves; pedicel articulation not
detectable in flower; hairs short, orange-
red, 0.2 mm long; androphore stellate-hairy 18. *C. mossambicensis* (p. 41)

23. Flowers single in each leaf axil; perianth
cream or brownish cream, divided for
$\frac{1}{3}$–$\frac{1}{2}$ its length into 4 stout fleshy, tightly
incurved perianth lobes; **T** 3 .... 14. *C. usambarensis* (p. 37)
Flowers clustered in fascicles in each axil;
perianth greenish or yellowish, divided for
$\frac{1}{5}$ its length or more into 5–9 flimsy, ± open
perianth lobes 24

24. Pedicels 0.6–0.9 mm diameter; perianth lobes
5–7 mm long; pedicels and outer perianth
with short, rusty red indumentum; **K** 4, 5,
T 3, 6–8 .... 19. *C. greenwayi* (p. 42)
Pedicels < 0.6 mm diameter; perianth lobes
3.5–4 mm long; fruiting carpels stipitate 25

25. Pedicel 0.2–0.3 mm diameter; perianth lobes
6–8(–10), reflexed; androphore (2.5–)
3–4 mm long; **K** 7, **T** 3, 6 .... 20. *C. pseudoclavata* (p. 43)
Pedicel 0.4–0.6 mm diameter; perianth lobes
5, ascending; androphore 0.2 mm long;
T 6–8 .... 21. *C. discoglypremnophylla* (p. 45)

Several West African species are cultivated in East Africa for their edible seeds ('cola nuts') which induce a feeling of alertness, well-being and contentment.

The most valued species are *C. nitida* and *C. acuminata*, both widely cultivated, medium-sized trees with white flowers ± 2.5 cm across, each perianth lobe having a red basal marking.

Cola acuminata (P. Beauv.) Schott & Endl.; U.O.P.Z.: 206 (1949); I.T.U: 24 (1953)
Cultivated in Uganda; Zanzibar. Tanzania, Muheza District: Zigi, Lunguza, *Shabani* 1064!
Cola nitida (Vent.) Schott & Endl.
Tanzania, Lushoto District: Amani, Sigi Shini Plantation, *Greenway* 2954!

Cola millenii K.Schum. is a small tree with 5-lobed, ± orbicular leaves, ± 10 cm long. The flowers are campanulate, pink to purple, ± 1 cm diameter. It is recorded in cultivation from near Nairobi, but is not known for its edible seeds (*K. Lennox* in EAH 10565!).

Cola verticillata (Thonn.) A.Chev.; T.T.C.L.: 594 (1949), produces seeds of lesser worth and is known from cultivation in Uganda (Mengo District: Kampala Plantation, *Snowden* 1966!) and Tanzania. It is readily recognized by the leaves (and inflorescences) arranged in whorls of three or four.

1. **Cola gigantea** *A.Chev.* in Bull. Soc. Bot. France: 55: 32 (1908); Brenan in K.B. 11: 151 (1956); Germain in F.C.B. 10: 315 (1963); Hamilton, Uganda For. Trees: 117 (1981). Types: Central African Republic, Bondjos, *Chevalier* 5152 (P, syn., K photo.!); Forêt de Possel, *Chevalier* 11184 (P, syn.); Dar Banda Oriental, Mbélé, Gounda, *Chevalier* 7300 (P, syn.; K!, isosyn.)

Deciduous tree 10–35 m tall, with spreading crown; bole usually long and straight, sometimes with small, short buttresses; bark thick, rough, grey or brown, with deep or slight vertical fissures; slash fibrous, pink to red or mottled yellow, reticulated, turning darker; no exudate recorded; ultimate branchlets 5–8 mm thick, densely rusty tomentose with stellate hairs 0.2–0.5 mm across; bud-scales soon falling, triangular, ± 8 mm long, 8 mm wide, rusty tomentose. Leaf-blade ovate, 15–37(–60) cm long, 10–32(–45) cm wide, apex rounded, base cordate, the internal angle obtuse, ± 120°, rarely truncate, leathery, digitately 5–11-nerved, drying brown or greenish brown above, mature leaves glossy, glabrous, white and densely tomentose beneath; leaf-blades of immature plants larger, ± 45 cm long, 45 cm wide, 3–5-lobed, divided for up to $^1/_4$, subglabrous; petiole terete, 3.5–22 cm long, 0.2–0.35 cm thick, slightly swollen at base and apex, rusty tomentose when young; stipules caducous. Inflorescences borne with the leaves (often on short spur shoots that only produce leaves after flowering), 2–4 per stem, 3–12 cm long, 2–5 cm wide, densely rusty tomentose with stellate hairs; bracts triangular, 1.5–2 mm long; ± 5 branches from the main axis, lowest branch 0.4–3.5 cm long, with 2–4, 1–3-flowered partial peduncle; pedicels ± 1.5 mm. Flowers with perianth white becoming pink, sweetly scented, campanulate, 7–8 mm long, 5.5–10 mm across divided for $^1/_3$ or more into 5 acute, reflexed lobes 2.5–3.5 mm long, 2 mm wide, the margin conspicuously involute; outer surface densely reddish tomentose with minute 8–13-armed stellate hairs, inner surface glabrous in the lower 2.5–3 mm, the upper parts thickly white tomentose with 5–7-armed sinuous stellate hairs. Male flowers with androphore 0.7–0.8 mm long, 0.7 mm wide at the base, tapering strongly to the apex, glabrous, but surrounded at the base by long stellate hairs, anther thecae in 2 distinct whorls of 5–7, forming a cylinder 1.5 mm high, 1.5 mm wide, the upper thecae with stellate hairs; vestigial ovary 0.3 mm long, glabrous. Female flowers unknown. Fruit with follicles obliquely ovoid-ellipsoid, ± 20 cm long, 17 cm high, 6 cm wide, rostrum short and stout, ± 0.7 cm long, stipe absent, thickly covered in minute rusty stellate hairs; seeds oblong-ellipsoid ± 3.5 cm long, 2 cm wide, sessile.

UGANDA. West Nile District: Metuli, Dec. 1932, *Eggeling* 889!; Bunyoro District: road to Butiaba, no date, *Dawe* 779!; Mengo District: Kasala, Nov. 1914, *Dummer* 1288!
DISTR. U 1, 2, 4; Ghana, Togo, Benin, Nigeria, Cameroon, Central African Republic, Congo-Kinshasa, Sudan
HAB. Gallery forest and secondary semi-deciduous forest; 950–1500 m

SYN. *C. cordifolia* sensu I.T.U., ed. 1: 234 (1940) & ed. 2: 415, photo. 36 (1952), *non* (Cav.) R.Br.

LOCAL USES. Hamilton (*loc. cit.*) records that Chimpanzees eat the fruit walls and the 'arils'.
CONSERVATION NOTES. Although known from only six specimens in Uganda, this species has such a large range that it is considered of "least concern" for conservation.

2. **Cola pierlotii** *Germain* in B.J.B.B. 32: 495 (1962) & in F.C.B. 10: 315 (1963). Type: Congo-Kinshasa, Mt Kahusi, Tshinganda Forest, *Pierlot* 3334 (BR, holo.)

Evergreen tree 10–15 m high; slash slimy, without exudates; ultimate branchlets ± 4 mm thick, dull dark brown, glabrous; axillary buds subwoody, globose, 3 mm diameter, apex with spine 2 mm long; bud scales not seen, soon falling. Leaves alternate, or, less usually, subopposite; blade elliptic, (10–)15–20 cm long, (4–)6–8.5 cm wide, acumen slender, 1.2–1.8 cm long, base rounded-obtuse, leathery, drying brown, lateral nerves 7 pairs, midrib raised above, upper and lower surfaces with thinly scattered appressed, white, 8-armed stellate hairs, ± 0.2–0.3 mm diameter; petiole terete, longitudinally grooved when dry, 1–7 cm long, 1.5–2 mm thick, slightly swollen at base and apex, glabrous; stipules caducous. Inflorescence borne on leafy stems below the leaves, 1 per stem, racemose, ± 3.5 cm long and wide, often with a branch at base, 6–7-flowered, densely white stellate tomentose, with hairs as on leaves, extending to the outer perianth; bracts caducous; pedicels ± 9 mm, articulated 4 mm from the perianth. Flower colour not reported, campanulate, ± 13 mm long and wide, divided into 5 very thick, triangular lobes, ± 5 mm long and wide, outer surface with indumentum as inflorescence, inner with the median third appressed stellate-hairy, with bands of hairs extending to the ovary, otherwise glabrous. Male flowers not seen. Female flowers with indehiscent anthers scattered at base of ovary; ovary oblong in side view, ± 3 mm long, 3.5 mm wide, densely puberulous, carpels 5, stigmas, sessile, flat, fleshy, smooth, appressed to top of carpels, forming a cap ± 1 mm deep, 4 mm wide. Fruit with 5 follicles developing, follicles patent, subcylindrical, ± 14 cm long, 3.5 cm wide, including a laterally compressed, slightly twisted rostrum 3 cm long, 1.2 cm deep, 0.7 cm wide, and a 1.5 cm long, 1 cm wide stipe; seeds not seen, but reported red (*Hafashimana* 727).

UGANDA. Kigezi District: Bwindi forest, Ishasha Gorge, fr. Nov. 1997, *Hafashimana* 490! & Ihihizo, fl. Aug. 1998, *Hafashimana* 727!
DISTR. **U** 2; eastern Congo-Kinshasa
HAB. Evergreen forest; 1350–1550 m

LOCAL USES. None are known in Uganda.
CONSERVATION NOTES. Although known in E Africa only from the two specimens cited above, *C. pierlotii* is known from 31 specimens in Congo-Kinshasa where it is confined to the District Lacs Edouard et Kivu between 1200–2000 m altitude (Germain 1962 loc. cit.). The number of specimens known suggests that within its range, it is fairly common so it is here assessed as "least concern". However, clearance of forest in this area would pose a threat and necessitate reassessment.
NOTES. Ishasha Gorge has been known for decades as a site where many Congolian species have their only locality in the FTEA area. It is remarkable that this species has remained undiscovered there so long.

3. **Cola scheffleri** *K.Schum.* in E.J. 33: 314 (1903); T.T.C.L.: 593 (1949); Brenan in K.B. 11: 163 (1956). Type: Tanzania, Lushoto District: Usambara Mts, Derema, *Scheffler* 150 (B†, holo.; BM, EA, iso., K photo.!)

Evergreen tree 25–30 m tall; trunk 30–60 cm diameter with smooth grey or brown bark and longitudinal fissures, crown pyramidal or oblong; ultimate branchlets 5–7 mm, reddish brown, with brown stellate hairs when young; bud-scales caducous, subulate, ± 9 mm long, 1 mm wide, apical bud thickly golden brown felty-tomentose. Leaf-blade entire, oblong 7.5–25 cm long, 4–13 cm wide or ± orbicular in outline and digitately 3(–5)-lobed, 12–32 cm long, 13–30(–40) cm wide, divided for ²/₃, the lateral lobes shorter, apex acuminate, acumen up to 1.7 cm, base rounded (subcordate in juvenile leaves), papery, glabrous above and below; petiole red, terete, 4–9 cm long, 1.5 mm wide, with brown tomentum when young; stipules caducous. Inflorescences 4–5 per stem, borne amongst the leaves, paniculate, 3.5–6 cm long, 2.5–10 cm wide, thickly covered in felty reddish brown hairs, lowest branch 0.7–1.6 cm from the base, 0.7–1 cm long, bearing 2–3 partial peduncle, each 1–2 mm long, bearing a single flower; bracts caducous, not seen; pedicels 1–2 mm long. Flowers pinkish brown outside, red with whitish pimples inside, perianth campanulate to widely obconical, 12–20 mm long, 14–20 mm wide, divided for ± half its length into 5–6 strap-shaped, reflexed lobes, outside with brown-black, felty tomentum of 5–9-armed stellate hairs, densest at the base of the perianth, inner surface glabrous apart from the involute margins of the lobes. Male flowers with androphore tapering slightly towards the apex, 5–7(–10) mm long, 1 mm wide, puberulous, with inconspicuous, mostly simple white hairs; anthers 8–10, uniseriate, forming a short cylinder 1.2–2 mm long, 3.5–4 mm wide, affixed to the head of the androphore by a glabrous disc; vestigial carpels ± 0.5 mm long, glabrous, largely concealed in a flask-like cavity in the head of the androphore. Female flowers known only from the type, slightly larger than the male, carpels 5 mm long, tomentose. Fruit with five bright red or brown patent follicles, each subglobose and laterally flattened, ± 6 × 5 × 3 cm, apex subrostrate, stipe 0.3–0.5 cm long, 1 cm wide, glabrous, 3–5-seeded, the seeds embedded in jelly-like pulp; seeds oblong-ellipsoid, ± 2.5 cm long, 1.2 cm wide.

TANZANIA. Lushoto/Tanga District: below Longuza Hill, Sigi R., fl. 19 Nov. 1947, *Brenan & Greenway* 8347!; Tanga District: Mlinga, sterile 18 Feb. 1937, *Greenway* 4910!; Morogoro District: Kanga, fr. Mar. 1989, *Manktelow et al.* 89/232!
DISTR. **T** 3, 6, 7; endemic to Tanzania
HAB. Evergreen forest; 650–1500 m

LOCAL USES. Seed pulp edible (*Pocs* 6136B).
CONSERVATION NOTES. This species, although known from 16 specimens, distributed among seven sites, several of which are at least nominally protected, seems relatively common. Luke (in litt.) reports this as a locally common pioneer of medium sized semi-deciduous forest gaps (16–100 sq. m). In the absence of information that its forest habitat at these sites is in danger, this species is here treated as "near threatened".
     IUCN (Red List 2002, www.redlist.org) list this species as vulnerable (VU B1+2b), however, this is on the basis of "Occurring only in the Nguru Mts and in the south of the Udzungwa Mountains at Kihanzi". This seems unjustified in view of its actual, much larger, range.
NOTES. Apart from the type, only one other flowering collection has been made. Most of the 16 specimens known are sterile. Identification then depends on the characteristically 3–5-lobed leaf and the golden brown, densely felty-tomentose apical bud. Confusion with e.g. *Sterculia appendiculata*, also with lobed leaves, is possible.
     While two of the fertile specimens have only entire leaves, those of the other specimens are mostly or always ± lobed. It seems likely that juvenile trees and suckers of this species develop deeply lobed leaves, whilst smaller, entire leaves are present in flowering branches of mature trees (more difficult to collect), much as in *Cola gigantea* (q.v.). This heterophylly has caused confusion (Brenan, K. B. 1: 143 (1956)). Although known from two disjunct areas (eastern Usambaras and Kanga Mt, near Lushuto), no morphological differences have been detected between the two. However, flowers are unknown from the Kanga Mt area.
     Brenan (loc.cit.) ascribed *Sterculia lindensis* to *Cola scheffleri* with some doubt. However Dorr has shown that this is not a *Cola*, but a *Sterculia*.
     A 17th specimen was reported while this account was in press: *Luke et al.* 8187, Udzungwa Mountains NP (Luke pers. comm.), extending the range to **T** 7.

4. **Cola stelechantha** *Brenan* in K.B. 11: 143 (1956). Type: Tanzania, Morogoro District: Turiani, *Paulo* 182 (K!, holo., NHT, iso.)

Evergreen shrub or small tree 3–10 m tall; branches dark greyish brown, young stems with scattered, appressed greyish black stellate hairs; bud-scales caducous, filiform, ± 5 mm long, 0.5 mm wide, tomentose, brown. Leaf-blade obovate to elliptic, 5.5–28 cm long, 2.5–15.5 cm wide, acumen ± 1 cm long, base cuneate, ± 5 pairs of main veins, glabrous above and beneath; petiole terete, ± 13 cm long, 0.15–0.3 cm thick, swollen, black and kneed at base and apex, indumentum as stem. Inflorescence cauliflorous, in old leaf axils of branches 2–3 cm wide, ± sessile, clustered in large numbers, forming woody burrs 3 cm wide, 1 cm long; bracts 3–5, bilobed, distichous, increasing in size from the base up, up to 2.5 mm long, 1.5 mm wide; pedicels (6–)10–20(–28) mm long, 1 mm wide, articulated 2–4 mm from the base, indumentum greyish brown, stellate. Flowers deep pink to dark red, drying blackish purple, ± 14 mm across, perianth divided to within 1–2 mm of the base into 4–6 patent lobes, lobes ovate-triangular, 5.5–7 mm long, 4 mm wide, margins involute, membranous, running full length of lobe, undulate, white, ± 1.2 mm wide, the outer surface of the lobes thinly covered with 4–16-armed stellate hairs, the largest hairs partly scale-like (rarely glabrous); inner surface thickly covered in papillae sometimes with occasional or thinly scattered red stellate hairs. Male flowers with androphore cylindrical, 2–3 mm long, 0.8 mm across, glabrous or with scattered stellate hairs; anthers uniseriate, 8–10, glabrous. Female flowers as the male, but androphore absent, non-functional anthers forming a ring at the base of the ovary; ovary (3–)5-lobed, very densely grey, stellately tomentose, ± ovoid, 2–2.5 mm long, 1.5–2 mm wide; stigmas ± sessile, reflexed, ± 1 mm across, black. Fruit borne ± 1 m from the ground on a woody, hooked pedicel 1.5–2 cm long, 0.3 cm thick, follicles 5, patent, green or yellow marked red, each ellipsoid, ± 6–7 cm long, 3 cm wide, rostrum 1.5 cm long, sessile, glabrous, slightly warty, particularly in distal portion, 6–12-seeded; seeds red, angular, ± 9–11 mm diameter.

TANZANIA. Iringa District: forest block to E of Udekwa village, fl. Dec. 1981, *Rodgers & Hall* 2267! & Udzungwa Mts, Nyumbenito area, fl. Dec. 1981, *Rodgers & Hall* 1474!; & Udzungwa Mts National Park, Mt Luhombero Pt 129, fl. 25 Sep. 2000, *Luke et al.* 6623!
DISTR. **T** 7; not known elsewhere
HAB. Evergreen forest; 900–2100 m

LOCAL USES. None are known.
CONSERVATION NOTES. *C. stelechantha* is known from 14 specimens and ± seven sites, most of which are forest reserves and therefore, protected to an extent. It seems restricted to the Udzungwa Mts where it is relatively common between 900–2000 m. Forest in this area is threatened by wood extraction (Frimodt-Moeller pers. comm.), so the species is here assessed as VU B2a, b(iii), i.e., vulnerable.
NOTES. Very closely related (the "*C. stelechantha* group") to *C. porphyrantha* Brenan, *C. rondoensis* Cheek, *C. quentinii* Cheek and *C. lukei* Cheek, but distinct in the characters stated in the key. These five species are unique in East Africa in bearing flowers on the trunk and main branches, rather than bearing axillary flowers, also in that the perianth in the male flower is held flat, rather than being campanulate, finally in that the stigmas in the female flowers are white, papillate, reflexed, and ± obovate (where known).
    In describing *C. lukei* as a new species, I erred in describing the range of variation in outer perianth indumentum of *C. stelechantha*, and in stating that it had 3–4 carpels. It usually has 5.

5. **Cola porphyrantha** *Brenan* in K.B. 40: 85 (1985); K.T.S.L.: 163 (1994). Type: Kenya, Kwale District: Shimba Hills National Park, Longo Mwagandi Forest, *Brenan et al.* 14557 (K!, holo.; BR!, C!, EA, FT!, G!, K!, M!, MO!, P!, S! iso.)

Evergreen tree 7–20 m tall; stems glabrous; bud-scales very early caducous, ovate, ± 3 mm long, 3 mm wide. Leaf-blade elliptic, rarely slightly ovate or oblanceolate, 8–25 cm long, 2.5–10 cm wide, acumen up to 1.5 cm long, base cuneate to subcordate, glabrous, 5–6 pairs of veins; petiole terete, very variable in length, 0.5–13 cm long, 0.1–0.4 mm thick, glabrous, drying black at base and apex; stipules caducous, lanceolate, acuminate, ± 7 mm long, 1.5 mm wide, margin pubescent. Inflorescence cauliflorous, fascicles of single flowers from short woody spur shoots ± 0.7–1.5 cm long, 0.7–1 cm wide each bearing 15–30 flowers, isolated or clustered in burrs 4–5 cm across, on both branches and trunk, bracts basal, 3–4, ± 1.1 mm long, 0.8 mm wide, rounded; pedicel 1.3–4 cm long, articulated 0.5–1 cm from the base, glabrous, bright red. Flowers bright red or red-purple, rose-purple or (*Luke* 1835) yellow to flesh-coloured inside, 12–16 mm across, perianth divided to within 1.5–3.5 mm of the base into 4–6 patent lobes, lobes ovate-triangular 3.5–9 mm long, 3.5–5 mm wide, often revolute, glabrous outside, minutely papillate inside. Male flowers with white androphore, 1.5–2.5 mm long, 0.3–0.5 mm wide, glabrous, anthers 5–7, uniseriate, forming a ring 0.5–1 mm long, 1.2–1.5 mm thick, glabrous. Female flowers as the male, but androphore absent, non-functional anthers forming a ring at the base of the ovary; ovary 3–4 lobed, very densely grey, stellately tomentose, ovoid, 1.2–2 mm long, 1.2–1.5 mm wide; stigmas ± sessile, reflexed, ± 1 mm wide. Fruits with carpels reddish, ovoid-round, ± 2–3 cm long, 1.5–2 cm high, 1–1.5 cm thick, apex shortly rostrate, glabrescent. Fig. 3, p. 29.

KENYA. Kilifi District: Pangani Rocks, fl. 1 May 1989, *Luke* 1835!; Kwale District: Shimba Hills National Park, Longo Mwagandi Forest, 16 May 1968, *Magogo & Glover* 1080 & idem, 18 Nov. 1978, *Brenan et al.* 14557!
TANZANIA. Lushoto District: E Usambara Mts, 12 km NE of Amani, Bulwa, fr. Feb. 2001, *Cordeiro* 160! & idem, fl. 25 Oct. 2003, *Ndangalasi* 801! 803!
DISTR. **K** 7; **T** 3; not known elsewhere.
HAB. Evergreen forest, often on limestone; 45–950 m

LOCAL USES. The fruits are edible (*Magogo & Glover* 1080).
CONSERVATION NOTES. The seven specimens available suggest that this spectacular species is known for certain from only two small forest sites (Longo Mwagandi in Shimba Hills and Pangani Rocks) in Kenya and one in Tanzania, at Bulwa. Protection at the first site is currently of concern due to a conflict in management (Luke pers. comm.). The Tanzanian site consists of less than 5 individuals in a 12.6 ha disturbed forest fragment near Bulwa in which this is the only *Cola* (Cordeiro pers. comm. 2001–2003). Although disjunct altitudinally (by ± 500 m) and geographically from the two Kenyan sites, careful comparison of the complete material obtained through Cordeiro reveals no significant morphological differences. *Cola porphyrantha* is here assessed as EN B2a,b(iii) i.e. Endangered. IUCN (Red List 2002, www.redlist.org) also list this species as endangered (EN B1+2c).
NOTES. *Cola porphyrantha* falls in the " *C. stelechantha* group" (see notes under the last). The species are easily separated using the characters in the key.

6. **Cola lukei** *Cheek* in K.B. 57: 417 (2002). Type: Tanzania, Lushoto District: E Usambara Mts, Kwamgumi Forest Reserve, *Luke & Muir* 6105 (K!, holo.; DSM, EA, NHT, iso.)

Monoecious much-branched tree 6–10 m tall; branches greyish brown, with numerous shallow furrows, lenticels inconspicuous; leafy stems 2–3 mm diameter, internodes 1–20 mm long, the distal ± 10 cm of actively growing stems brown-scurfy with dense stellate hairs. Leaf-blade simple, elliptic, 11–32 × 4.5–10(–13.5) cm, entire, acumen short and wide, 0.2–1.3 cm long, margin barely revolute, base acute, with 6–7 pairs of dull yellow secondary nerves; petiole terete, 0.5–12.5 cm long, both short and long petiole on the same stem, pulvinate at base and apex, the pulvini (2–)6–9(–12) mm long, ± 0.2 cm diameter, brownish green below, brown-scurfy as the distal part of the stems; stipules caducous, not persisting below the apical bud, triangular-acuminate, 4–5 × 1.5–2 mm, glabrescent. Inflorescences cauliflorous, loosely fasciculate, with several flowers opening at one time from a mass of 20–30 buds on one of several woody burrs (brachyblasts) scattered on the trunk; pedicel 3.7–9.5 mm long, articulated 3–4(–6) mm from the base, bearing 1–3 scattered caducous bracts below the point of articulation, indumentum minutely puberulous, of patent simple hairs with scattered 5–10-armed stellate hairs; bracts suborbicular, ± 1.7 × 1.5 mm, apex bifid. Female flowers 20–30 mm diameter, pale yellow; perianth divided into 4–5 patent or reflexed lobes, subtriangular to narrowly ovate, 9–12 × 7–8 mm, apex acute to obtuse, base ± 5–8 mm wide, margin involute, membranous, inconspicuous; basal part of perianth campanulate, ± 5 × 5 mm, outer perianth puberulous, as the pedicel, inner surface glabrous, basal half of lobes minutely papillate; anthers 8–9, uniseriate, indehiscent, at the base of the ovary; carpels 3–4, adnate, ovoid, 5 × 4 mm, puberulous; stigmas white, elliptic, 3.5–4.5 × 2–2.5 mm, reflexed, apex rounded, upper surface papillate. Male flowers as the female, but 14–17(–22) mm diameter, perianth lobes 5–6, 5.5–7 × 3.2–4.5 mm, basal part of perianth slightly convex, ± 6 mm wide; androphore cylindrical, 1.5–3 × 0.5–0.7 mm, apex slightly constricted, glabrous; anther-head (2.5–)3 mm wide, anthers uniseriate, 8–9; carpels inconspicuous. Fruit one or more per stem; pedicel accrescent, stout, 5 mm long; fruitlets 3, red, patent, widely ellipsoid, ± 4.5 × 3.2 cm, ± 2 cm deep, apex

FIG. 3. *COLA PORPHYRANTHA* — **1**, leafy shoot × ²/₃; **2**, inflorescence × ²/₃; **3**, male flower × 2; **4**, androecium × 10; **5**, ovary of female flower showing basal staminodes × 10; **6**, young carpels × ²/₃; 7, fruiting carpel × ²/₃. 1–6 from *Brenan et al.* 14577, **7** from *Magogo & Glover* 1080. Reproduced from Kew Bull. 40: 86. Drawn by Eleanor Catherine.

rostrate, 6–10 × 2–4 mm, base sessile, ventral suture ridge conspicuous, pericarp leathery, ± 2 mm thick when dried, smooth, densely shortly dark brown puberulous with stellate and simple hairs; seeds ± 7 per fruit, hemispherical, ± 15 × 15 mm, invested in a thick glossy seed-coat.

TANZANIA. Lushuto District: East Usambara Mts, Kwamgumi Forest Reserve, fl. 13 Dec. 1999, *Luke & Muir* 6105! & idem, fr. 13 Dec. 1999, *Luke & Muir* 6107A!
DISTR. **T** 3; not known elsewhere
HAB. Lowland forest with *Pterocarpus*; 300-350 m

LOCAL USES. Used as firewood.
CONSERVATION NOTES. *Cola lukei* was assessed as Endangered (EN B1+ 2c, C2b) in Cheek (Kew Bull. 57: 417, 2002). The species is only known from the eastern Usambara Mts at Kwamgumi Forest Reserve.

7. **Cola rondoensis** *Cheek* **sp. nov.** a *C. quentinii* Cheek floribus in quoque fasciculo numerosis non tantum 1–2, pagina interiori perianthii dense hirsuto non glabro, androphoro glabro non stellato-piloso differt. Typus: Tanzania, Lindi District: Rondo Forest Reserve, *Bidgood, Abdallah & Vollesen* 1554 (K!, holo.)

Cauliflorous shrub 3 m tall; stems 5 mm diameter, glossy brown, with numerous raised, smooth, darker brown, longitudinally elliptic lenticels, glabrous; bud-scales not seen. Leaf-blade coriaceous, glossy, obovate-oblong (larger leaves) or elliptic (smaller leaves), 6.5–26 cm long, 3–11.5 cm wide, acumen 1 cm long, base obtuse or rounded, lateral nerves 5–7 on each side of the midrib, glabrous; petiole terete, 3–75 mm long, 2 mm wide, glabrous; stipules not seen, caducous. Inflorescence cauliflorous, forming woody burrs on the trunk, fasciculate, with numerous flowers in bud, only 1–2 flowers open at one time; bracts 4 per flower, entirely sheathing each flower stalk, ovate, lower bracts ± as wide as long, 1–2 mm long, 1.5–2 mm wide, deeply sessile, upper bracts longer than wide, ± 4 mm long, 1 mm wide, acute; pedicel 3–4 mm long, 1 mm wide, articulated 1 mm from the base, indumentum greyish brown, stellate. Male flowers red, ± 16 mm across, perianth flat, divided to within 2–3 mm of the base into 5 patent lobes, lobes ovate-triangular, 5–6 mm long, 3.5–4 mm wide, margin involute (distal three-quarters) membranous, straight, not undulate, concolorous, inconspicuous; outer and inner lobe surfaces densely white stellate-hairy; androphore cylindrical, 1.2–2 mm long, 0.5 mm wide; anthers uniseriate, ± 7, anther head 1 mm long, 2 mm wide, glabrous. Female flowers and fruit unknown.

TANZANIA. Lindi District: Rondo Plateau, Rondo Forest Reserve, fl. 14 Feb. 1991, *Bidgood, Abdallah & Vollesen* 1554!
DISTR. **T** 8; not known elsewhere
HAB. Evergreen forest with *Milicia*, *Albizia* and *Dialium*; ± 650 m

LOCAL USES. None are known.
CONSERVATION NOTES. The only known site for this species is an amphitheatre-like patch of forest 1–2 ha in extent that is fed by several streams that drain the Rondo Plateau above. About half a dozen trees were seen when the type collection was made. No immediate threats to this forest patch were noted, although some forest on the plateau above has been converted to plantation and forest degradation due to pit-sawing is a possibility (Bidgood and Vollesen pers. comm. 2005). Data on the forests of the Rondo Plateau can be found in Bidgood & Vollesen in K.B. 47: 759–764 (1992) and also in Vollesen in Davis et al. 1994, Centres of Plant Diversity 1: 225–226. WWF & IUCN. *Cola rondoensis* is here assessed on the basis of the data above, as critically endangered CR B2a, b (iii).
NOTES. Within the *Cola stelechantha* group (see notes under that species), *C. rondoensis* is most likely to be confused with *C. quentinii* (on limestone at Kimboza, **T** 6) and *C. lukei* (E Usambara Mts) since all three have short flower stalks.

8. **Cola quentinii** *Cheek* **sp. nov.** a *C. lukei* Cheek perianthio rubro non luteolo, androphoro dense hirsuto non glabro, carpellis et stigmatibus 5 non 3–4 differt. Typus: Tanzania, Morogoro District: Uluguru Mts, Kimboza Forest Reserve, *Luke et al.* 7624 (K! spirit, holo.; BR, EA, K!, NHT, iso.)

Cauliflorous tree 5–15 m tall; stems probably as *C. stelechantha*. Leaf-blade oblong-elliptic, 14–52 × 7.5–27 cm, acumen ± 1.5 cm, base obtuse, abruptly and shortly concave-cuneate, lateral nerves 4–5 on each side of the midrib; petiole 0.9–27 cm long, 2–4 mm wide, basal pulvinus 3–8 mm wide, glabrous. Inflorescence cauliflorous, in old leaf axils of stems 1–2 cm diameter, forming 1–4-flowered woody burrs up to 1.5 cm diameter; bracts single, caducous, bilobed, concave, ± 3 mm long and wide; stalks 3 mm long (0–2 mm long in female flowers), articulated 0–0.5 mm from the base, ovoid lateral bud immediately below the articulation, indumentum dense, stellate, hairs 5–7-armed. Male flowers red, speckled green, 14 mm across, perianth flat, divided to within ± 3 mm of the base into 5–6 lobes, lobes ovate-triangular, lobes 4–5 mm long, 4 mm wide, apex slightly acuminate, margin involute in distal half, membranous, ± 0.3 mm wide, straight, concolorous, inner surface thickly covered in papillae, lacking hairs, outer surface as flower stalk; androphore cylindrical, ± 2.5 mm long, 1.3 mm wide, densely stellate-hairy, anthers uniseriate, ± 10 mm long, 3–3.5 mm wide, glabrous. Female flowers as the male but ± 18 mm across (to 30 mm in *Mhoro* 320, when pressed), widely campanulate, perianth lobes ± 9 mm long, 7 mm wide, ovary ovoid-conical, 4 mm diameter, shortly stellate-hairy, stigmas white, 5, 6 mm across, each obovate, ± 3 mm wide, reflexed, papillate. Fruits unknown. Fig. 4, p. 32.

TANZANIA. Morogoro District: Uluguru Mts, Kimboza Forest Reserve, fl. 4 Nov. 1987, *Luke* 772 & idem, pt 204, fl. 19 Sep. 2001, *Luke et al.* 7624! & Mkungwe Forest Reserve, fl. 8 Oct. 2000, *Mhoro* UMBCP 320!
DISTR. **T** 6; not known elsewhere
HAB. Forest on limestone; 200–500 m

LOCAL USES. None are known.
CONSERVATION NOTES. *Cola quentinii* was only known from the ± 385 ha Kimboza Forest until the recent collection by Mhoro, made after the account for this species was written. At least ten other taxa of flowering plant appear to be strictly restricted to the Kimboza forest. These taxa are: Annonaceae genus indet. A of FTEA, *Impatiens cinnabarina*, *Garcinia bifasciculata*, *Ligelsheimia silvestris*, *Necepsia castaneifolia* subsp. *kimbozensis*, *Baphia pauloi*, *Cyphostemma* sp. 'P' of FTEA, *Turraea kimbozensis*, *Streptocarpus kimbozanus* and *Cola kimbozensis* (Luke pers. comm., edited by Vollesen, 2005). Kimboza forest is notable, apart from the high number of strictly endemic taxa, for being (a) the only forest area of any size left on the lower (below 1500 m) slopes of the Uluguru Mts; (b) perhaps the best example of Tanzanian wet forest on Jurassic karstic limestone; (c) the richest of all East African lowland forests for avifauna (71 forest species of birds). This forest, although of very high conservation value, is gazetted as a production forest reserve, as much as 1% of the canopy being removed for timber each year, under licence. Other pressures are pole-cutting and charcoal burning, according to Rodgers et al. (1983), The Conservation Values and Status of Kimboza Forest Reserve, Tanzania, from which most of the foregoing is taken. Quentin Luke (pers. comm., 2005) reports that between 1987 and 2001 there was noticeable degradation of the forest by humans. He considers the most serious impact, however, to be from the invasion of the exotic timber species, *Cedrela odorata* from the adjoining plantations. Careful selective removal of this species is a priority if the conservation value of Kimboza is to be maintained. A further threat is the destruction of the limestone substrate to produce building material, he reports.
   Cola quentinii is here assessed as CRB2a,b(iii) that is, critically endangered in view of the data cited above on its geographic range and habitat threats. Luke reports that in 1987 on a fleeting visit, he saw only a single tree. During his second visit in 2001 he found 3 or 4 trees. The species is probably rare even within Kimboza forest given that, despite the visits of numerous able botanists only two collections are known of the taxon, by one botanist. Given that it is very likely that less than 250 mature individuals of the species survive it can also be assessed as critically endangered on the basis of its estimated small population size (CR C1 + C2a(ii)).

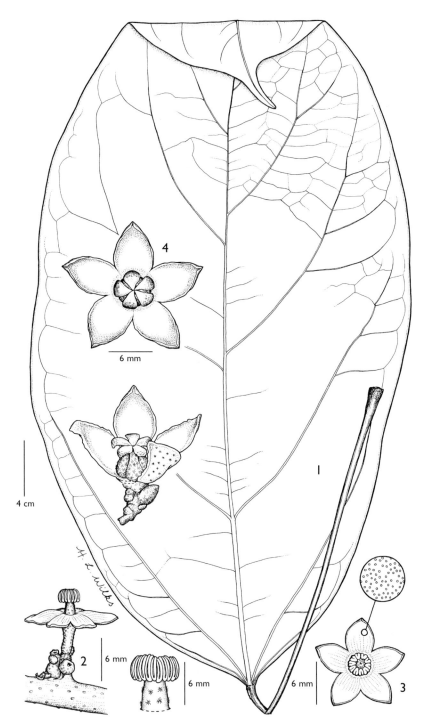

FIG. 4. *COLA QUENTINII* — **1**, leaf, upper surface; **2**, inflorescence with male flower, side view, showing detail of androgynophore and anther head; **3**, male flower, in plan view with detail of tepal papillae; **4**, female inflorescence and flower, side view, with one tepal removed (below) and intact, in plan view (above). All from *Luke et al.* 7624. Drawn by Hazel Wilks.

NOTES. Since the name *Cola kimbozensis* is already employed, *Cola quentinii* was chosen to honour Quentin Luke, the Kenyan botanist who had made the only known collections of the taxon and who first raised the possibility that they might be new to science.

The affinities of *Cola quentinii*, which falls in the *C. stelechantha* group (see notes under that species) are with *C. lukei* and *C. rondoensis*, both of which also have short-stalked flowers with poorly developed membranous margins to the perianth lobes. It differs from the first in the characters mentioned in the diagnosis, and from the second in that its bracts appear caducous (not persistent and sheathing), in the stellate-hairy androphore and in the glabrous inner surface to the perianth lobes (not glabrous and stellate-hairy respectively).

More collections are needed to complete our knowledge of the species.

9. **Cola octoloboides** *Brenan* in K.B. 33: 283 (1978); K.T.S.L.: 163 (1994). Type: Kenya, Kilifi District: Chasimba, *Adams* 124 (K!, holo.; EA iso.)

Evergreen shrub or tree 3–6 m tall; trunk pale, stems 4–5 mm thick, whitish; current year's growth pale brown, velvety tomentose, glabrous at length; bud-scales ovate, 4–5.5 mm long, 1.5–3 mm wide, acuminate, often persistent. Leaf-blade ovate (smallest leaves) to obovate-elliptic (larger), 3–18(–34) cm long, 1.8–7.5(–15) cm wide, acumen 0.5–0.7 mm long, 0.3–0.4 mm wide, base cordate (smallest leaves) cuneate to rounded, veins prominent below, 6–10 pairs, glabrous above and beneath; petiole terete, 0–4(–6.3) cm long, 1.5 mm wide, the base and apex distinctly swollen and 2.5–3.5 mm wide, thickly covered in a pale brown, velvety tomentum, becoming glabrous; stipules linear, rarely narrowly rhombic, 6–8 mm long, 0.2–0.5 mm wide, tomentose, ± persistent. Inflorescence with flowers borne with the leaves, axillary, single or rarely in fascicles, sessile; bracts ± 4, ± orbicular, the lowest ± 3 mm long, 4 mm wide, increasing to 6 mm long and wide, outer surface tomentose, inner glossy, glabrous, dark brown; pedicel absent. Flowers opening yellowish, fading chocolate brown, widely campanulate, 15–19 mm long, ± 28 mm across, perianth fleshy-leathery, divided within 2–5 mm of the base into 6 lobes, lobes narrowly triangular, 10–14 mm long, 4–9 mm wide, the margin folded inwards, outer surface densely stellate-hairy, with a mixture of long, lanate, colourless hairs up to 2 mm long, and smaller stellate hairs, folded margin lacking longer hairs, innner surface with minute papillae. Male flowers with androphore abruptly contracted from the base, 2–4 mm long, 3–3.5 mm across (4 mm across at the base), thickly long stellately hairy; anthers 14, uniseriate, forming a ring 2 mm long, 5.5–6 mm wide, inner surface tomentose. Female flowers without androphore; anthers at the base of the ovary; ovary subspherical, 3–4 mm long, 5–6 mm wide, densely stellate tomentose; style up to 0.5 mm long; stigmas 6, 0.5 mm long, partly reflexed, black. Fruit (probably immmature) with carpels felty green-white, 2.5 cm long, 1.3–1.8 cm wide, rostrum 0.5 cm long, stipe 0.8 mm long.

KENYA. Kilifi District: Kaloleni–Kilifi, 14 km, Chasimba, fl. 30 Dec. 1970, *Faden et al.* 70/943! & idem, 22 km, fl. 12 Dec. 1974, *Adams* 125!; Kwale District: Dzombo Hill, y.fr. 9 Feb. 1989, *Robertson et al.* MDE 301!
DISTR. **K** 7; not known elsewhere
HAB. Evergreen forest on limestone; 200–400 m

LOCAL USES. None are known.
CONSERVATION NOTES. Apart from the specimen from Kwale District, this species appears restricted to the forest remnants on the limestone at Chasimba. Its total area of occupancy is believed to be below the 500 km² threshold. Only five collections are known. The main threat at Chasimba is the removal of trees for firewood and a plan to use the rock to upgrade the Kilifi to Mariakani road (Luke, pers. comm.). *Cola octoloboides* is here assessed as EN B2a,b(iii), i.e. Endangered. IUCN (Red List 2002) also list this species as endangered (EN B1+2c).
NOTES. *Cola octoloboides* is still imperfectly known: fruiting material has not been seen. The persistent bud-scales and stipules together with the persistent stem indumentum make this one of the few East African *Cola* species easily recognized in the sterile state. It is most closely related to *C. ruawaensis* (see diagnosis of that species for distinguishing characters).

10. **Cola ruawaensis** *Cheek* **sp. nov.** a *C. octoloboide* Brenan stipulis caducis non persistentibus, bracteis profunde retusis non rotundatis, perianthio diviso usque ad 7–8 mm nec 2–5 mm e basi, androphoro pilos simplices non stellatos ferenti differt. Typus: Tanzania, Lindi District: Ruawa forest, *Mbago et al.* 2279 (K!, holo.; DAR, iso.)

Shrub 5 m tall; leafy stems white, longitudinally ridged, 3 mm wide, lenticels inconspicuous; current year's growth pale brown, tomentose, glabrous at length; bud-scales caducous, not seen. Leaf-blade simple, elliptic (smallest leaves) to obovate (larger), 7.5–16 cm long, 3.5–7 cm wide, acumen 0.3–1 cm long, base rounded to cordate (smallest leaves) or acute, lateral nerves 9 on each side of the midrib, prominent; petiole 0.5–6.5 cm long, 1 mm wide, the base and apex distinctly swollen, covered in pale brown tomentum, glabrescent; stipules caducous, not seen. Inflorescence with flowers borne with the leaves, axillary, in fascicles of 2–5, sessile, bracts ± 6 per flower and enveloping it in bud, papery, increasing in size with proximity to flower, innermost bract, suborbicular, ± 5 mm long, 8 mm wide, apex retuse, outer surface yellow-brown tomentose, apex, margins and inner surface glabrous, dark brown. Male flowers yellow, with brownish hairs, 20–25 mm across, perianth widely campanulate, divided to within 7–8 mm of the base into 6 or 7 lobes, lobes oblong-triangular, 12 mm long, 5 mm wide, apex obtuse, margin truncate, inner surface evenly papillate over the entire surface, outer surface brown tomentose, with two types of hair: short hairs dense, arms erect, long hairs scattered, arms appressed, 5; androphore conical-cylindrical, 5.5 mm long, 3.5 mm wide at base, 1.8 mm wide at apex, indumentum simple, patent, moderately dense; anthers uniseriate, 15; anther-head 2.5 mm long, 6.5 mm wide. Female flowers and fruits unknown. Fig. 5, p. 35.

TANZANIA. Lindi District: Ruawa Forest, fl. 12 Dec. 2001, *Mbago et al.* 2279!
DISTR. **T** 8; not known elsewhere
HAB. Lowland evergreen forest dominated by *Scorodophloeus* on coral rag; ± 330 m

CONSERVATION NOTES. *Cola ruawaensis* is only known from the type locality. Ruawa forest is threatened by agricultural encroachment and illegal logging according to Clarke, P. (1995) Status Reports for 6 Coastal Forests in Lindi Region, Tanzania. Accordingly the taxon is here assessed as CR B2a,b(iii), critically endangered.
NOTES. *Cola ruawaensis* is closely related to *C. octoloboides* of Kenya, also restricted to limestone, and only known from 1 or 2 sites. It differs in the non-persistent stipules, the deeply retuse (not rounded) bracts, the perianth divided to 7–8 mm from the base (not 2–5 mm), and the androphore with simple hairs (not stellate hairs). More collections are needed to complete our knowledge of the species.

11. **Cola congolana** *De Wild. & T. Durand* in B.S.B.B. 38(2): 181 (1899); Germain in F.C.B. 10: 292 (1963). Type: Congo-Kinshasa, between Matende and Kibala, *Dewèvre* s.n. (BR, holo.)

Evergreen, laxly branched tree 3–10 m high; bark silvery, brownish grey, flaking in thin papery scales, inner bark pale cream, fibrous, sapwood cream, heartwood darker; ultimate stems 3–4 mm thick, silvery grey, with numerous fine ridges, tomentose when very young, often bearing at length epiphytic liverworts and mosses; bud-scales early caducous, narrowly ovate, ± 8.5 mm long, 3 mm wide. Leaf-blade narrowly elliptic to oblanceolate 7.5–24 cm long, 3–9.5 cm wide, the apex with acumen ± 1.5 cm long, 0.3 cm wide, base cuneate, veins prominent above and below, sparsely tomentose below when young, becoming glabrous on both sides; petiole finely ridged, 0.7–6.5 cm long, 0.1–0.2 cm wide, thinly tomentose when young; stipules caducous. Inflorescences borne below the leaves, with fascicles of 1–20 single flowers per axil, few open at one time, stalks 5–8 mm long, articulated ± ¹/₃ the way from the base, subglabrous; bracts 4–7, increasing in size towards the apex, wider than long, 1.5–7.5 mm long, 2.5–9.5 mm wide, bilobed for up to half their length, the lowest bracts with rounded lobes, the uppermost acute, becoming perianth-like.

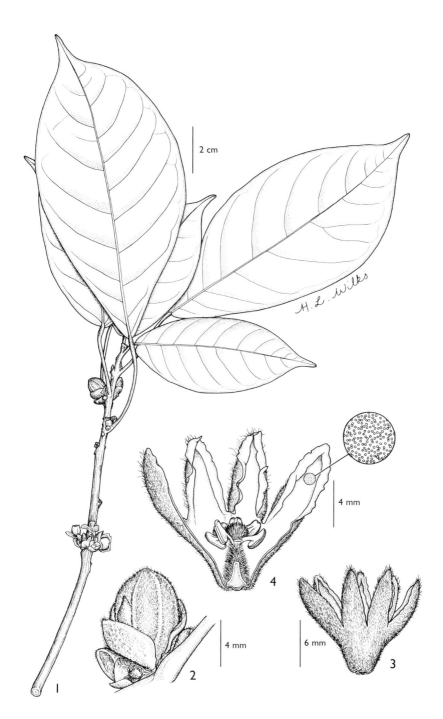

FIG. 5. *COLA RUAWAENSIS* — **1**, habit; **2**, flower buds with bracts; **3**, perianth, side view; **4**, male flower, sublongitudinal section. All from *Mbago* 2279. Drawn by Hazel Wilks.

Flowers deep cream, suffused with pink, sometimes yellow or orange, perianth campanulate, 10–14 mm long, 12–18 mm across, divided into (4–)5(–6) triangular lobes each 7–9 mm long, 4–5 mm wide, outside with appressed reddish brown stellate hairs, increasing in density towards the base, and spreading to the apical 1–2 mm of the inner surface, inner surface otherwise glabrous, glossy, with numerous papillae. Male flowers with androphore ± cylindrical, 3 mm long, 1 mm wide, glabrous; anthers 8–10, uniseriate, forming a cylinder 1.5 mm long, 2.5–3 mm wide, vestigial carpels 0.6 mm long, glabrous. Female flowers usually slightly larger, perianth often 6-lobed, ovary sessile, with a ring of stamens at the base, widely ovoid, 3–4 mm long, 3–4 mm wide, densely villose; style stout, ± 1 mm long; stigmas drying black, elongate, 1–1.5 mm long. Fruits with follicles irregularly long ellipsoid to slightly obovoid, 4–10 cm long, 2 cm wide, rostrum 0.5–1 cm long, 0.5 mm wide, stipe absent, glabrous, crimson; seeds spherical to ellipsoid, 1–1.3 cm long.

UGANDA. Toro District: Kibale, Hima Forest, 1 km Mpanga R. bridge–Kampala, 22 Dec. 1949, *Hoyle* 1403! & Itwara Forest, 29 Jan. 1945, *Greenway & Eggeling* 7053!; Ankole District: Kalinzu Forest, July 1938, *Eggeling* 3764!
DISTR. U 2; Congo-Kinshasa
HAB. Evergreen forest; ± 1500 m

SYN. *C. bracteata* De Wild. in Pl. Bequaert. 1: 522 (1922); I.T.U.: 415 (1952); Brenan in K.B. 11: 150 (1956); Hamilton, Uganda For. Trees: 118 (1981). Type: Congo-Kinshasa, Avakubi, *Bequaert* 1836 (BR, holo.)

LOCAL USES. None are known.
CONSERVATION NOTES. Although known in East Africa only from four specimens, *C. congolana* is relatively widespread and abundant in the forests of Congo-Kinshasa and so is considered "least concern" for conservation. IUCN (Red List 2000, www.redlist.org) list this species under its synonym *C. bracteata* De Wild., as VU B1+2c on the fallacious basis that it is endemic to Uganda.

12. **Cola uloloma** *Brenan* in K.B. 11: 150 (1956); K.T.S.L.: 163 (1994). Type: Tanzania, Pangani District: Bushiri Estate, *Faulkner* 654 (K!, holo.)

Evergreen small tree, rarely a shrub, (1.5–)3–10(–25) m tall, with a clean trunk and dense, rounded crown; bark pale grey, rough or smooth with longitudinal striations; ultimate branchlets 1 mm thick, dark brown, with numerous smooth raised lenticels of the same colour; bud-scales caducous. Leaf-blade narrowly elliptic to oblanceolate, sometimes slightly asymmetrical, 4–12 cm long, 1.7–4.7 cm wide, acumen 0.5–1.8 cm long, ± 0.3 cm wide, leathery, 5–7 pairs veins, glabrous above and beneath; petiole terete, 2–5 mm long, 1 mm wide, glabrescent; stipules caducous. Inflorescence in leaf-axils, with fascicles of 1–5 single flowers; bracts 3–4, ± orbicular, the lowest largest, 1.7–2.5 mm long, 2–3 mm across, deeply bilobed, drying with longitudinal ridges, glabrous; pedicel 5–11 mm long, articulated ± midway, puberulous. Flowers whitish, yellowish or pinkish brown, aromatic, 6–9.5 mm across, divided to within 1.5–2 mm of the base, lobes narrowly triangular, (2.5–)3–4 mm long, 1.2–1.5 mm wide, the upper margins reflexed, white, outer surface sparsely brown stellate-hairy, the hairs 0.1–0.2 mm diameter, inner surface minutely papillate. Male flowers with androphore 1.5–2 mm long, 0.2 mm wide, glabrous, anthers 5, uniseriate, forming a ring 1.5–1.8 mm long, 0.4–1.3 mm wide, glabrous. Female flowers as the male, but androphore absent, anthers forming a ring at the base of the ovary; ovary of 3–5 carpels, ± spherical, 1.2–2 mm long, 1.8–2.5 mm wide, stigmas ± 0.5 mm long, reflexed. Fruits with a single carpel developed, obliquely obovoid, ± 25 mm long, 17 mm wide, stipe indistinct, rostrum absent, orange, subglabrous.

KENYA. Kilifi District: Kaya Kambe, fr. 9 July 1987, *Robertson & Luke* 4787!; Kwale District: Shimba Hills, Mwele Mdogo Forest, y.fr. 23 Aug. 1953, *Drummond & Hemsley* 3972! & Pangani Rocks, fl. 10 July 1987, *Luke & Robertson* 494!
TANZANIA. Pangani District: Mwera, Kwa Besa, Mwanamgaru, fl. 28 March 1957, *Tanner* 3462! & Bushiri estate, *Faulkner* 654!

DISTR. **K** 7; **T** 3, ?6; not known elsewhere
HAB. Coastal forest; 30–500 m

LOCAL USES. None are recorded.

CONSERVATION NOTES. Of the 13 specimens seen, only five fertile gatherings, listed above, have been seen for *Cola uloloma*: this species seems rare, poorly known and restricted to a few small coastal forest fragments, mostly the Kaya forests of Kenya. Its extent of occurrence is estimated as less than 20,000 km². Given that forest quality is declining in several of these patches, and that less than 10 locations are known, *Cola uloloma* is here assessed as VU B2a,b(iii), i.e vulnerable to extinction.

NOTES. The uniformly short petiole (less than 7 mm) seen in *Cola uloloma* and *C. chlorantha* (q.v. for diagnostic characteristics) is unique in East African members of the genus and make these two species easily recognized in the sterile state, assuming that they are recognized as *Cola* in the first place. Most other species of *Cola* have a mixture of very long-stalked (to 19 cm) and very short-stalked leaves.

Hawthorne 1645 from the Pugu Hills (**T** 6), though sterile, probably represents this species, but without flowers or altitudinal data it is difficult to be certain. Luke (pers. comm.) reported sight records in Kenya for the lower (Kaya Muhaka) and upper (Mangea Hill) altitudinal ranges cited above. Luke (pers. comm.) also reported seeing what was possibly this species, but might also be *C. chlorantha* (especially the more southern and higher altitudinal specimens) in **T** 6 (Kimboza, *Luke* 8814 EA n.v.) and also in **T**7 (*Luke* 8747 EA n.v., 900m alt. and *Luke* 11311 EA n.v., 1330m alt.).

13. **Cola chlorantha** *F.White* in B.J.B.B. 60: 83 (1990). Type: Malawi, N Viphya Plateau, Choma Forest, *Dowsett-Lemaire* 471 (K!, holo.; FHO, iso.)

Evergreen tree ± 5 m high, closely resembling *C. uloloma* Brenan (see above) and indistinguishable vegetatively, differing in the inconspicuous bracts, more slender pedicels (± 0.5 mm thick, not 0.2–0.3 mm thick), flowers green (not whitish, pinkish or yellowish brown), the bowl-shaped perianth tube, ± 3.5 mm diameter at apex (not cup-shaped, ± 2–2.3 mm diam at apex) and the ascending, not patent, perianth lobes, anthers 7, not 5, carpels 6–7 not 3–5. The fruits and seeds of *C. chlorantha* are unknown.

TANZANIA. Kigoma District: Kasakati, fl. Sep. 1965, *Suzuki* B-45!; Iringa District: Mufindi, Lulanda, fl. 24 Nov. 1988, *Gereau & Lovett* 2559!
DISTR. **T** 4, 7; Malawi
HAB. Evergreen forest; ± 1500 m

LOCAL USES. None are recorded.

CONSERVATION NOTES. *Cola chlorantha* is only known from three widely separated sites, at which a decline in habitat is inferred. It is here assessed as VU B2a,b(iii), i.e. vulnerable, in view of the data above, and the area of occupancy being estimated at less than 2,000 km². More observations are needed on the status of this species at the sites indicated.

14. **Cola usambarensis** *Engl.* in E.J. 39: 595 (1907); Brenan in B.J.B.B. 18: 4 (1946); T.T.C.L.: 594 (1949); Brenan in K.B. 11: 151 (1956). Type: Tanzania, Lushoto District: Amani, *Engler* 3423 (B†, holo.)

Evergreen tree up to 15 m high; stems terete, 1.5–4 mm wide, brown or grey, sparsely covered with appressed grey stellate hairs when young, soon becoming glabrous; bud-scales falling early, ovate-triangular, 1–3 mm long, ± 1 mm wide. Leaves often with spherical woody galls ± 5 mm diameter on the petiole, less often on the blade; blade oblanceolate to narrowly elliptic, the smallest ovate to elliptic, 2.5–25 cm long, 1.2–10(–13) cm wide, acumen blunt, 6–12 mm long, 5 mm wide, base cuneate, 5–6 pairs of main veins, glabrous above and beneath; petiole terete, 4–10 cm long, the smallest leaves subsessile, indumentum as stem; stipules linear, 6.5 mm long, 1 mm wide. Inflorescence in leaf-axils at least 3–4 leaves below the stem apex, 1(–2)-flowered; bracts 4–6, increasing in size towards the flower, the largest bract conspicuously bilobed, ± 1 mm long, 1.3 mm wide, brown, glabrous; stalk stout, 1–2 mm long, indumentum as perianth. Flowers with perianth cream or brownish cream, ± campanulate, divided for $\frac{1}{3}$–$\frac{1}{2}$ into 4 incurved lobes; outer surface stellate-pubescent, inner surface thickly covered in papillae. Male flowers 4.7–6 mm long; androphore 1.5 mm long, 0.4 mm wide at the base tapering to 0.3 mm at the apex, glabrous; anthers uniseriate, 8–10, glabrous, in a disc ± 1 mm long, 1.5 mm diameter; ovary vestigial, concealed. Female flowers 7.5–10 mm long; androphore absent; anthers reduced, 7 or 8 at base of ovary; ovary globose, 3–3.8 mm long, 4.2–4.5 mm wide, densely and coarsely tomentose; style up to 0.8 mm long; stigmas 4, black, patent, 2 mm long. Fruits with up to 4 carpels developing, carpels ascending, shortly oblong-cylindric, up to 5 cm long, 3 cm wide, 2.5 cm deep, rostrum triangular, laterally flattened, 5–7 mm long, 3–5 mm wide, base sessile, suture with a low ridge; densely stellate-hairy with minute powdery golden brown hairs, pericarp papery-leathery; seeds in two ranks, up to 14, testa white, cotyledons red (*fide Luke & Muir* 6086).

TANZANIA. Lushoto District: Amani-Zigi Plot, fl. fr. 12 Dec. 1999, *Luke & Muir* 6088! & Amani, 21 Dec. 1928, *Greenway* 1069! & Amani, base of Mt Bomole, 24 Dec. 1956, *Verdcourt* 1726!
DISTR. **T** 3; not known elsewhere
HAB. Evergreen forest; ± 900 m

LOCAL USES. Used for poles for house construction and medicine (Muir loc. cit.).
CONSERVATION NOTES. *Cola usambarensis* is listed as DD (Data Deficient) by IUCN (2002 Redlist, www.redlist.org). A study to investigate the factors affecting the distribution of *Cola usambarensis* was concluded as an MSc Thesis by Muir (1998, Univ. College, London). Although detailed data were gathered, confusion as to the identification of the species arose. Material assumed to be *Cola usambarensis* at some of the study sites later proved to be an unrelated species, *C. lukei* (Cheek in K.B. 57: 417 (2002)). Although Muir (1998) records the species from six protected areas, there is only evidence in the form of fertile specimens to support the existence of *C. usambarensis* at one of these, the Amani Forest Reserve, in the East Usambara Mountains, which occupies less than 100 km². Muir (loc. cit.) records that the species is harvested for use as poles, medicine, and in one case, rope (from the bark). While there is some doubt as to which, if not all, these species of *Cola* these uses apply to, there is no doubt that *C. usambarensis* is at risk and that its habitat is declining due to the illegal harvest of building materials (Muir 1998). It is here assessed as Critically Endangered (CR B1a+b(iii)), reflecting that it is only known from a single location, that there is a decline in habitat quality and that its extent of occurrence is less than 100 km². Muir (op. cit.) rates the species as Endangered.
NOTES. Only three flowering specimens, cited above, are known to me. In the absence of flowers, identification is uncertain, although the 5 mm diameter globose woody galls present on the leaves may be unique to this species. However, not all specimens seem to bear them. Brenan (1956, *op. cit.*) also mentions *Zimmerman* 6759 (EA - buds) from Kwamkoro, 12 Nov. 1907. Ruffo *et al.* (in Hamilton and Bensted-Smith, Forest Conservation in the East Usambara Mountains, Tanzania. IUCN (1989)) mention two further collections: *Ruffo & Mmari* 2025 (Kwamkoro F.R. & Amani area) and *Ruffo & Mmari* 2246 (Kwamsambia/Kihuhwi F.R.), probably at Lushoto and/or EA: these have not been verified by me. The only fruiting collections known are *Luke* 5245, 6080, 6085 and 6086, all from Mbomole Hill.

15. **Cola minor** *Brenan* in K.B. 11: 149 (1956); K.T.S.L.: 162, map (1994). Type: Kenya, Kwale District: Marenji Forest, *Dale* in F.H. 3786 (K!, holo.; EA, iso.)

Evergreen shrub or tree 3–12 m tall, usually single-stemmed; bark and wood unknown, sap colourless; ultimate branchlets terete, 0.7–1.5 mm wide, greyish brown, pubescent, becoming puberulous; bud-scales caducous, triangular, ± 2 mm long, puberulous. Leaf-blade oblanceolate, rarely narrowly elliptic, 1.9–10(–12) cm long, 0.7–3.6(–4.5) cm wide, apex bluntly acuminate, the acumen ± 1 cm long, base cuneate to acute with convex sides, (7–)8–9 pairs of main veins, leathery, glabrous, dark green drying grey green; petiole terete, 1.5–17(–32) mm long, ± 1 mm wide, puberulous to scurfily brown or grey pubescent when young. Inflorescences axillary, usually amongst the leaves but also on older wood below, fasciculate with numerous single flowers in bud, but only 1–3 flowers open at one time; bracts 2–3, rounded to oblong, hooded, often bilobed at apex, 1–2.5 mm long, to 1 mm wide, pubescent; stalk 3–4.5(–8) mm long, 0.3–0.4(–0.8) mm wide, with stiff reddish stellate hairs; articulated ± a third of its length from the base. Flowers white, greenish or yellow, fragrant, perianth divided for ⁴/₅ into (4–)5 partly reflexed lobes, each 2.2–3(–4) mm long, 1.7–2.5(–3) mm wide, outer surface as pedicel, inner surface with finer hairs on upper half of lobes and minutely papillate. Male flowers with androphore 1.2–2.2 mm long, 0.1–0.2 mm wide, glabrous; anthers 8–10, uniseriate, glabrous, in a disc 0.5–0.7 mm long, ± 1 mm diameter; ovary vestigial, concealed. Female flowers sometimes rather larger than the male, with androphore absent; anthers barely reduced at the base of the ovary; ovary subglobose, slightly trilobed, 1.5–2 mm long, 2.5–3 mm wide, densely tomentose; style ± 1.5 mm long, stigmas 3, recurved. Fruits with 1(–3) fruitlets developing, fruitlets obliquely clavate, 8–17 mm long, 10–13 mm wide, apiculus absent, stipe ± 1.5 mm long, orange-yellow, pubescent, with accrescent calyx.

KENYA. Kilifi District: Watamu Coast, fl. 18 Jan. 1957, *Langridge* 1/57!; Kwale District: Mrima Hill Forest, fr. 4 Sep. 1957, *Verdcourt* 1863! & Jombo Mt, fl. 26 Sep. 1982, *Polhill & Robertson* 4830!
TANZANIA. Tanga District: Pangani, Madanga, Bushiri, fl. 7 May 1957, *Tanner* 3509!
DISTR. **K** 7; **T** 3; not known elsewhere
HAB. Coastal evergreen forest or thicket; 50–250 m

LOCAL NAMES & USES. None are known.
CONSERVATION NOTES. Of the 12 specimens known to me of this species, only one is from Tanzania, the remainder being from **K** 7. It is limited to the severely fragmented remnants of coastal forest, now threatened by clearance. I estimate the area of occurrence of this species to be less than 2000 km². Accordingly, *C. minor* is here assessed as VU B2a, b(iii), that is vulnerable.
NOTES. Sometimes confused with *Cola pseudoclavata* which shares the same habitat and, occasionally, location. *C. minor* is distinguishable from *C. pseudoclavata* by its constantly shorter petioles (up to 17(–32) mm long) and dark green uniformly smaller (to 10(–12) cm long) leaf-blades with 8–9 pairs of nerves. All specimens of *C. pseudoclavata* have petioles up to 50(–60) mm long and yellowish-green leaf-blades with a greater range in size (to 17 cm long) and 5–7(–8) pairs of nerves. *C. minor* is only known from 10 fertile collections. On present evidence it appears dioecious.

16. **Cola kimbozensis** *Cheek* **sp. nov.** a *C. minore* Brenan laminis foliorum 9–16 cm (non 1.9–10 cm) longis, nervis lateralibus utroque costae latere (4–)5–6 (non (7–)8–9); pedicellis 8–9 mm (non 3–4 mm) longis; lobis perianthii glabris non hirsutis intus differt. Typus: Tanzania, Morogoro District: Kimboza Forest Reserve, *Rodgers, Hall & Mwasumbi* 2624 (K!, holo.; DAR, iso., 2 sheets)

Evergreen tree, 4–7 m tall; bark and wood unknown; ramiflorous, flowering stems leafless, ± 3 mm diameter, white, finely longitudinally ridged, glabrous, the short (0.5–2 cm long) leafy stems either lateral (spur shoots) or terminal, green, ± 1 mm diameter, smooth, glabrous, each bearing (1–)2–3 leaves during the flowering season. Leaf-blade obovate (2.5–)9–16 cm long, (1.8–)3.8–6.5 cm wide, acumen slender, 7 mm long, edge of cuneate base concave, lateral nerves (4–)5–6 on each side of the midrib, leathery, drying bright grey-green, glabrous; petiole terete, very sparsely hairy, (2–)9–18 mm long, drying green, strongly swollen and black at base and apex, often angled at apex, glabrous; stipules caducous, not seen. Inflorescences 1-several cm below the leaves, fasciculate, forming woody burrs ± 3 mm long, with numerous single flowers in bud, but only 1–3 mature flowers present at one time; bracts not seen; stalk 8–9(–12) mm long, 0.2–0.3 mm wide, articulated 1–3 mm from the apex, sparsely hairy. Flowers cream-coloured, fragrance not recorded, perianth divided into 5 lobes, each ± 2.5 mm long, inner surface glabrous, outer surface sparsely pubescent, androphore glabrous. Female flowers unknown. Fruits and seeds unknown.

TANZANIA. Morogoro District: Uluguru Mts, Kimboza Forest Reserve, fl. Oct. 1983, *Rodgers, Hall & Mwasumbi* 2624! & near Ruvu bridge, fl. 4 Nov. 1987, *Luke* 771! & Kimboza, 13 June 2002, ster., *Luke & Luke* 8821
DISTR. **T** 6; not known elsewhere
HAB. Evergreen forest on limestone with *Scorodophloeus*; 200–450 m

Local uses. None are known.

Conservation notes. *Cola kimbozensis* is only known from the three collections cited from the Kimboza forest, notes upon which are given under *Cola quentinii*. According to *Rodgers et al.* 2624, it is common there, at least locally, but despite this, due to its very narrow geographical range and the threats to its habitat (see under *Cola quentinii* above), it is here assessed as CR B2a, b(iii), that is critically endangered.

17. **Cola microcarpa** *Brenan* in K.B. 11: 147 (1956). Type: Tanzania, Morogoro District: Turiani, *Semsei* 1466 (K,!, holo.; EA iso.)

Evergreen tree, 5–18 m tall; bark and wood unknown; ultimate branchlets terete, 2–3 mm wide, with reddish brown scurf when young, soon or at length lost, revealing a smooth white epidermis; bud-scales triangular, falling early. Leaf-blade oblanceolate or elliptic, 2–15(–17.5) cm long, 0.8–6.8 cm wide, apex acuminate, base acute to cuneate, 9–10 pairs of main veins, leathery, glabrous; petiole terete, 1–67 mm long, ± 1 mm wide, with reddish brown scurf often persisting; stipules ligulate, ± 6 mm long, puberulous. Inflorescences axillary, amongst the leaves, but also on older wood below, with fascicles of numerous single flowers, but only 1–4 flowers open at one time; bracts 2(–3), rounded to oblong, ± 1.5–2 mm long, 1.5 mm wide, apex often bilobed, puberulous; flower stalk 9–23 mm long, articulated ± a third its length from the base, with long, shaggy grey hairs. Flowers yellowish green, perianth divided into 5(–7) lobes each 4.5–8 mm long, 2.2–4.3 mm across, outer surface as stalk, rarely slightly yellowish, inner sparsely hairy in upper half and densely and minutely papillate throughout. Male flowers with androphore terete, ± 3 mm long, glabrous; anthers uniseriate, 5–6, glabrous, in a disc 1–1.5 mm long, 1.5–2.5 mm diameter; ovary vestigial, concealed. Female flowers with androphore absent; anthers barely reduced, at base of ovary; ovary subglobose, 1.1–1.8(–3) mm long; 2.2–4.5 mm diameter, densely yellowish, long-tomentose; style 0.7–1 mm long; stigmas (3–)4, recurved, ± 0.7 mm long, with stout hairs. Fruitlets peach-coloured, downy, subglobose 1–1.4 cm long, 1.1–1.8 cm wide, stipe ± 1 mm long, rostrum absent.

Tanzania. Morogoro District: Turiani, fl. Sep. 1953, *Eggeling* 6704! & idem, old fl. Nov. 1953, *Semsei* 1489!; Kilwa District: Kingupira, fl. 13 Jan. 1977, *Vollesen in MRC* 4323!
Distr. **T** 6–8; not known elsewhere
Hab. Riverine forest; 100–700 m

Local uses. None are known.

Conservation notes. Although a total of 15 specimens are known, very few of these are far from Turiani (**T** 6) in Morogoro District or Kingupira (**T** 8) in Kilwa District and no more than ten distinct sites are known in all, with an estimated area of occupancy of less than 2,000 km². Threats to the habitat of *Cola microcarpa* at some of these sites are documented in Clarke, Status Reports for 6 Coastal Forests in Lindi Region, Tanzania (1995), so this species is assessed as VU B2a, b(iii), i.e. "vulnerable". The only record for **T** 7 derives from Luke (pers. comm., not verified by me) who cites *Luke* 11445 from Udzungwa MNP, 530 m alt.

Notes. Brenan says of *C. microcarpa* that "*Wallace* 906 is said to occur at 1100 m in the Usambaras". The latter is very likely an error since this altitude is far too high for the species, and if it did occur there, it is likely that the intensive collecting efforts there in the 1980s would have produced more specimens. Specimens labelled as this species from **T** 4 (e.g. *Bidgood et al.* 2964 and 2963) represent another taxon that may be the closely related *C. mossambicensis* (flowers are required), differing e.g. in the absence of shaggy grey indumentum and in the higher altitudinal range (850–1200 m). Iversen records this taxon from E Usambaras according to Luke (pers. comm.) but I have seen no specimen and find this doubtful. Luke (pers. comm.) also reports this species from **T** 7 at Udzungwa MNP, *Luke* 11445 which is more credible.

　　Cola microcarpa seems rather more variable than most *Cola* species in flower shape and size. For example, pedicel width is 0.3 mm in the type collection, yet 1 mm in *Milne-Redhead & Taylor* 7349. Unfortunately Brenan, who uses this feature in his key to distinguish two groups of species including *C. microcarpa*, errs, firstly in placing the species amongst those with pedicels 0.75–1.5 mm in diameter (the opposing clause is "pedicels less than 0.5 mm in diameter") and secondly in his description of the species "circiter 0.5 mm diametro".

Palmer and Pitman (1972; 2: 1491) in *Trees of Southern Africa* seem first to mention that *C. microcarpa* extends to South Africa. They give an extensive description and illustration. Drummond in his check-list of Rhodesian woody plants (Kirkia 10: 260, 1975) lists *C. microcarpa* as a synonym of *C. greenwayi*, which is surprising since Wild in the Flora Zambesiaca account of *Cola* does not mention *C. microcarpa*. No collections have been seen of *C. microcarpa* from that region and it may be that Drummond was referring to *C. mossambicensis* (q.v.), material of which was once referred to as *C. microcarpa* to which it is closely related. Verdoorn in her paper 'The genus *Cola* in southern Africa' (Bothalia 13: 277, 1981) follows and supports Drummond's synonymy. However, I am convinced that the South African *C. microcarpa sensu* Palmer & Pitman (*C. greenwayi sensu* Verdoorn) is a related, but distinct and unnamed species.

Cola greenwayi can most reliably be differentiated from *C. microcarpa* by the short rusty-red indumentum of the outer perianth and pedicel (long, shaggy grey indumentum in *C. microcarpa*) and less consistently by the absence of thick, pale redddish scurf on young stems and petiole. *Cola greenwayi* occurs between 1100–1830 m altitude, *C. microcarpa* between 125–660 m.

18. **Cola mossambicensis** Wild in Bol. Soc. Brot. Ser. 2, 33: 39 (1959); Wild in F.Z. 1: 560 (1961); Palgrave, Trees of Southern Africa: 599 (1977). Type: Mozambique, Manica e Sofala, Espungabera, Gogoi Mt, *Torre* 4308 (LISC, holo., K!, SRGH, iso.)

Evergreen tree, (3–)18–35 m tall; bole fluted and buttressed in larger specimens, bark ashy grey or pale brown; ultimate branchlets terete, 4–5 mm wide, with reddish brown scurf when young, soon or at length lost, revealing a white, longitudinally ridged, epidermis; bud-scales not seen, caducous. Leaf-blade oblanceolate or elliptic, (2–)14–21 cm long, (0.8–)7–11 cm wide, apex acuminate, base acute to rounded, 12–15 pairs of main veins, quaternary veins conspicuous on both surfaces, leathery, glabrous; petiole terete, 1–100 mm long, 1–2 mm wide, pulvini with caducous reddish brown scurf; stipules triangular, ± 3–4 mm long, 2–3 mm wide, grey pubescent, caducous, seen only in buds. Inflorescences axillary, amongst the leaves, with fascicles of ± 6–10 flowers; bracts numerous, concave, rounded-triangular, 2–3 mm long, 2 mm across, apex entire, outer surface densely brown pubescent; flower stalk 6–9 mm long, articulation absent or obscure in flower, basal in fruit, with dense orange-red hairs. Flowers white or cream, perianth divided for $^6/_7$ into 5 lobes each 4–7 mm long, 2–2.5 mm across, outer surface as stalk, inner densely hairy in upper half and densely and minutely papillate throughout. Male flowers with androphore terete ± 3 mm long, densely stellate-pubescent; anthers uniseriate, 5–6, glabrous, in a disc 1–1.5 mm long, 1.5–2.5 mm diameter; ovary vestigial, concealed. Female flowers with androphore absent; anthers barely reduced, at base of ovary; ovary subglobose, ± 1.5 mm diameter, densely pale brown tomentose; style 1.5 mm long; stigmas 4, recurved, ± 1 mm long, 0.7 mm wide, papillose. Fruitlets 1–5, brown, downy, transverse ridges faint, two pairs, subglobose-obovoid ± 1.4 cm long, 1.3 cm wide, stipe ± 2 mm long, rostrum ± 1.5 mm diameter.

TANZANIA. Lindi District: Rondo Plateau, Mtene, fr. March 1952, *Semsei* 733! DISTR. **T** ?4, 8; Malawi and Mozambique HAB. Forest; altitude unknown but probably less than 900 m

LOCAL USES. Mkupete (Kimwera), timber, fide *Semsei* 733.
CONSERVATION NOTES. *Cola mossambicensis* is listed as VU B1+2C, i.e. vulnerable, (www.redlist.org), based on an assessment by S. Bandeira in 1988 (Bandeira, S. 1995. Data collection forms for tree species of Mozambique). Bandeira noted that this species is severely threatened by the decline in quality and conversion of its habitat for agricultural purposes. However he cited the species as mainly occurring in central Mozambique, and only possibly occurring in Malawi. In fact, of the 17 specimens known, most are from Malawi, where it occurs in Mlanje, Dedze, Ncheu and Zomba Districts, seven of the specimens occurring in Zomba District. Six specimens are known from Mozambique, namely from Manica e Sofala, Zambezia and Zambezia-Niassa Provinces. Since the species is now known from more than ten sites (the 17 specimens are derived from 11 sites) and has an extent of occurrence that now exceeds the IUCN (2001), criterion for vulnerable status, it may be more appropriate to treat the species as near threatened (NT) unless conversion of its habitat can be shown to have been 30% or more in the last 100 years.

NOTES. *Cola mossambicensis* is only likely to be confused with *C. microcarpa*, since they both have the young stems and petiole covered in a thick scurfy-felt like layer of red-brown hairs that is shed to reveal white stems, unusual features in East African members of the genus. However, they differ from each other in several floral and fruit features (see the key). In addition, *C. mossambicensis* is distinctive in that it is a tall tree with a fluted bole, and in that the white stems are conspicuously longitudinally grooved, with hair remnants showing darkly in the grooves. The leafy stems are particularly straight, stout and bear leaves at unusually regular intervals (1.5–2 cm) for a *Cola*. Moreover the petioles are very long, 10 cm being common. These features make the species determinable with some confidence even when sterile, which is not normally the case for simple-leaved *Cola* species.

Specimens labelled *C. microcarpa* from **T** 4 (Kigoma District: Kasye Forest, buds 25 March 1994, *Bidgood et al.* 2962, ibid. fr. 2964; Mt Livandabe, y.fr.28 May 1997, *Bidgood et al.* 4152) may well represent *C. mossambicensis* since they have the distinctive characteristics referred to in the paragraph above. Their altitudinal range (850–1200 m) is also seen in the material of the species from Malawi. However flowering material is needed to confirm this identification. The limited fruiting material (only 3 mericarps are available) differs from *C. mossambicensis* in being more sparsely hairy than usual, and in lacking a rostrum and ridges, so it is also possible that this represents yet another new taxon for the area.

19. **Cola greenwayi** *Brenan* in K.B. 11: 144 (1956); Wild in F.Z. 1: 560 (1961); Germain in F.C.B. 10: 290 (1963); K.T.S.L.: 162, fig., map (1994). Type: Tanzania, Lushoto District: Mkusi [Mkuzi], *Greenway* 7891 (K!, holo.; EA, iso.)

Evergreen, much-branched tree with columnar or conical crown, 8–25 m tall; bark greyish-white or greyish-brown, reticulate and rough or smooth; ultimate branches terete, 2–2.5 mm wide, greyish brown, subscabrid with blackish hairs or glabrous; bud-scales caducous. Leaf-blade narrowly elliptic or oblanceolate, 3–16(–18) cm long, 1–6(–7) cm wide, apex long-acute, but sometimes shortly and bluntly acuminate, base acute to cuneate, 7–10 pairs of main veins, leathery, glabrous, dark green, drying brownish green; petiole terete, 0.3–4(–5.6) cm long, 0.1–0.15 mm wide, densely covered with short reddish or blackish subscabrid stellate hairs, at length glabrous; stipules caducous, narrowly triangular, 3–6 mm long, with brown, scurfy hairs. Inflorescences axillary, amongst the leaves, with fascicles of numerous single flowers, but with 1–6 flowers open at a time; bracts 1–3, basal, ± elliptic to triangular 1–1.5 mm long, 1.5–2 mm wide, sometimes inserted on the lower part of the pedicel and narrowly elliptic to deeply bifid, to 5 mm long, puberulous; pedicel 5–12(–15) mm long, 0.6–0.9 mm diameter, articulated $^1/_3$–$^1/_2$ the way from the base, with short felty rusty reddish hairs. Flowers yellowish brown or green, perianth divided almost to the base into 5 lobes, each (3.5–)5–7(–10.5) mm long, 2.2–3.5 mm wide, outer surface as pedicel, inner sparsely hairy and densely and minutely papillate. Male flowers with androphore terete (2.5–)3.5 mm long, ± 0.5 mm wide, glabrous or hairy; anthers uniseriate, 4–6, glabrous, in a disc ± 1 mm long, 1.5–2 mm diameter, ovary vestigial, concealed. Female flowers with androphore absent; anthers barely reduced at base of ovary; ovary subglobose (1.2–)2–3 mm long, 2–3.5 mm diameter, densely tomentose; style 1–1.5 mm long; stigmas (3–)4(–5), recurved, 0.5 mm long, with stout hairs. Fruits subglobose to clavate, 12–16(–25) mm long, 15–20 mm diameter, apiculus ± 1 mm long, stipe 1–2 mm long, reddish crimson or brownish orange, with reddish tomentum.

var. **greenwayi**

Perianth lobes (3.5–)5–7(–7.5) mm long; androphore glabrous.

TANZANIA. Lushoto district: Mkimbwe, Mkusi [Mkuzi], fl. 26 Dec. 1959, *Greenway* 9670!; Ulanga District: Mwanihana, Kilombero, fl. 25 Sep. 1981, *Rodgers & Hall* 1199/A!; Songea District: Luwira-Kitega Forest Reserve, fl. 19 Oct. 1956, *Semsei* 2533!
DISTR. **T** 3, 6–8; Congo-Kinshasa, Zambia, Malawi, Mozambique, Zimbabwe
HAB. Evergreen forest, often with *Podocarpus*; 1100–1850 m

SYN. *Cola sp.* of Brenan & A.P.D.Jones in B.J.B.B. 18: 4 (1946); Brenan in T.T.C.L.: 594 (1949), based on *Grant* H. 39/39!, H.39/40!

LOCAL USES. None recorded

CONSERVATION NOTES. The stronghold of this variety in E Africa is at Mkusi, Lushuto District: **T** 3. Collections from elsewhere in Tanzania are few and inconsistent with those from **T** 3. *Horlyk* in TZ 348 (Luhega F.R., **T** 6) has atypically large fruit; *Semsei* 2533 has subsessile flowers. Further material and studies are needed to show whether these represent populations which merit independent varietal status, or whether var. *greenwayi* is merely highly variable. Until this matter is resolved, this variety is here rated as "near threatened".

NOTES. Verdcourt (in litt.) and Luke (pers. comm.) note this taxon from the Taita Hills (*Faden* 71/975), Kasigau (*Faden* 69/463) and perhaps from Kilibasi (*Luke & Robertson* 2069) but I have not been able to confirm this. Neither have I been able to confirm Luke's record (pers. comm.) from **T** 7 (*Luke et al.* 7979, Udzungwa MNP) although this seems feasible.

var. **keniensis** *Brenan* in K.B. 11: 146 (1956). Type: Kenya, Nairobi Arboretum, *Gardner* 1193 (K!, holo.; FHO, iso.)

Perianth lobes 6.5–10.5 mm long; androphore densely stellate-hairy.

KENYA. Kiambu District: Limuru, fl. 21 June 1918, *Snowden* 632!; Nairobi District: Karura Forest, fr. 23 March 1957, *Bally* 11392! & Ruaraka, *Battiscombe* 1019!

DISTR. **K** 4, ?5; not known elsewhere

HAB. Evergreen forest; 1700–2150 m

SYN. *Cola sp.* of Battiscombe, Descr. Cat. Trees Kenya Col.: 30 (1926)

LOCAL USES. Timber tree, *Rammell* 1050.

CONSERVATION NOTES. Beentje (pers. com. 2006) reports the destruction of the forest sites of this taxon: Limuru forest has been destroyed, Karura is being destroyed, Ruaraka is now an industrial area. Historically locally common in, although confined to, **K** 4, with ten specimens known and an extent of occurrence of less than 5000 km², the taxon is here assessed as vulnerable to extinction: VU A2, B1a, b (iii). More detailed investigation may well reveal that this is conservative and that a higher threat rating is needed.

NOTES. *Dale* in K1048 from Mandaret, Sotik, represents what may be this taxon from **K** 5. However, it differs in having a larger fruit which is densely red hairy, and may represent a new taxon. Flowering material from this location is desirable. According to Luke (pers. comm.), *Luke & Robertson* 113 from **K** 6(**K** 4) – Emali Hill, may represent this taxon, but I cannot confirm this, since I have not seen the specimen.

Cola greenwayi is confined to high altitude forest. In flower it is conspicuous in the rust-red hairs that clothe the perianth. Various authors have confused this species with *C. microcarpa* (q.v.).

20. **Cola pseudoclavata** *Cheek* **sp. nov.** a *Cola clavata* Masters stipite fructuli male definiti a pedicello usque ad fructificationem leniter dilatanti, non bene definiti aequi latitudine per totam longitudinem; fructificatione ellipsoidea transversaliter non longitudinaliter; lobis perianthii extus pedicelloque pilos breviarmatos rubros stellatos, non longiarmatos albos ferentis differt. Type: Tanzania, Tanga District: Kivomila River, 8 km S of Ngomeni, *Drummond and Hemsley* 3519 (K!, sheet 2, holo.; EA, K! sheet 1, iso.)

Evergreen tree or shrub 5–21 m tall; bole sometimes buttressed at the base, bark smooth, pale grey or brown, wood hard, pale yellow or white; ultimate branchlets 1.5–3 mm thick, glabrous, smooth, whitish, becoming shallowly longitudinally ridged; bud-scales caducous. Leaf-blade ovate to oblanceolate, 2.5–17 cm long, 1.1–6.1 cm wide, apex acuminate, base cuneate to obtuse, sometimes rounded or slightly cordate, leathery, 5–8-nerved on each side of the midrib, finely reticulate, yellowish or lime green, drying brownish or greyish green, glabrous; petiole ± terete, 0.1–5(–6.5) cm long, ± 1 mm wide, slightly swollen at base and apex, glabrous; stipules caducous. Inflorescence axillary, amongst the leaves and on older wood below, fasciculate with numerous single flowers in bud, 1–3 flowers open at one time; bracts 3–6, ± elliptic, 1–2 mm long, entirely basal or inserted up to halfway along the pedicel, 1–2 mm long, increasing in size towards the flower, cucullate, entire or bilobed, puberulous; pedicel 4–6.5 mm long, 0.2–0.3 mm wide, with short-armed reddish stellate hairs; articulation basal, inconspicuous. Flowers with perianth cream inside, outside with brown hairs, scent not recorded, divided for $^1/_5$ into 6–8(–10) ± patent lobes, each (3.2–)3.5–4 mm long, 1–1.8 mm wide, the margins reflexed; outer surface as pedicel, inner with a few stellate hairs at tip, and minutely papillate. Male flowers with androphore (2.5–)3–4 mm long, 0.1–0.2 mm wide, glabrous, anthers uniseriate, 8–10, glabrous, in a disc ± 1 mm long, 1–1.5 mm diameter, cream or orange; ovary vestigial, concealed. Female flowers with androphore absent; anthers only slightly reduced at base of ovary; ovary depressed globose, 1–1.5 mm long, 1.5–2 mm diameter, strongly 5-lobed, bristly tomentose, cream; style 1–1.5 mm long. Fruitlets 1–3, ripening yellow and red, ± obliquely clavate, 1.4–1.9 cm long, 1.2–1.8 cm wide, apiculus oblique, inconspicuous, ± 1 mm long, covered in fine golden stellate hairs.

KENYA. Tana River District: Mnazini S Forest, fl. 15 June 1988, *Medley* 344!; Lamu District: Utwani Forest, fl. Sep. 1934, *Mohamed Abdullah* in F.D. 3353!; Kwale District: Mrima Hill Forest, fr. 4 Sep., 1957, *Verdcourt* 1855!
TANZANIA. Tanga District: Kivomila River, 8 km S Ngomeni, fl. 29 July 1953, *Drummond & Hemsley* 3519!; 3520!; Kilosa District: Vigunde, fr. Nov. 1952, *Semsei* 1017!
DISTR. **K** 7; **T** 3, 6; Somalia *fide* Thulin, Fl. Somal. 2: 37 (1999), based on sterile material (identification therefore uncertain).
HAB. Riverine or coastal evergreen or semi-deciduous forest; 30–600 m

SYN. *C. clavata* sensu Brenan in K.B. 11: 148 (1956), *non* Masters in F.T.A. 1: 222 (1868)

LOCAL USES. Wood used as handles for Jembes and axes, *Mahunda* 1. Choice firewood, *Medley* 344.
CONSERVATION NOTES. Known from ten fertile specimens in Kenya, four in Tanzania and numerous sterile specimens (identification therefore uncertain), *C. pseudoclavata* is fairly common within a fairly large range and appears not to be of immediate concern for conservation, although threats to its forest habitat could change this. For the moment it is rated as "near threatened".
NOTES. Dioecious on the evidence of *Drummond & Hemsley* 3519 & 3520. The former specimen is the only one known to have female flowers. *Cola pseudoclavata* is sometimes confused with *C. minor* in coastal Kenya, which is distinguished vegetatively by its smaller, dark green leaves with more uniformly short leaf-stalks. Sometimes confused in Tanzania with *C. discoglypremnophylla* (for differences, see there).

　　　Hitherto known in the FTEA area as *C. clavata* following Brenan (loc. cit. 1958). The type of *C. clavata* is a fruiting collection from Mozambique. Brenan, in attempting to work out the correct name for this FTEA taxon, (then known only from flowering material), concluded that it was *C. clavata* on the basis of his observations of perianth fragments persisting at the base of the fruits on the Mozambiquan type. However Wild (in litt.) in working out the *Cola* species of the Flora Zambesiaca region doubted Brenan's conclusion and postulated that the FTEA taxon differed and required a new name. In the intervening 40 years ample specimens have now accumulated of what are clearly two morphologically distinct and geographically disjunct species. These are easily distinguished using the characters in the table below. Accordingly a new name has been provided for the FTEA taxon above.

	Cola clavata	*Cola pseudoclavata*
Fruitlet width	0.9–1 cm	1.2–1.8 cm
Seed orientation in fruitlet	Longitudinal	Transverse
Fruitlet stipe	Well-defined, ± 10 mm long, width even, 2 mm	Poorly defined, ± 7 mm long, dilating steadily from pedicel to fruit body, ± 4 mm wide at midpoint
Perianth lobe number	5	6–8(–10)
Indumentum of outer perianth & pedicel	Long-armed white stellate hairs	Short-armed red stellate hairs
Geographical distribution	Mozambique	**K** 7, **T** 3 & 6

21. **Cola discoglypremnophylla** *Brenan & A.P.D.Jones* in B.J.B.B. 18: 2 (1946); T.T.C.L.: 593 (1949); Brenan in K.B. 11: 150 (1956); Wild in F.Z. 1: 561 (1961). Type: Tanzania, Lindi District: Lake Lutamba, *Schlieben* 5433 (BM, holo.; BR, K!, iso.)

Evergreen shrub or tree 3–10 m tall; bark smooth with dark and pale grey patches; ultimate branchlets terete, 2–4 mm wide, whitish, glabrous, becoming longitudinally ridged; bud-scales caducous. Leaf-blade ovate to obovate, 2–17 cm long, 0.8–10 cm wide, apex obtuse to slightly acuminate, base subacute to rounded, rounded, rarely subcordate, 4–5(–6) pairs of main veins, veinlets highly reticulate, leathery, glabrous, dark green, drying grey green; petiole terete, 0.2–5.5(–10.5) cm long, 1–1.4 mm wide, glabrous; stipules caducous, narrowly triangular, ± 3 mm long, soon falling. Inflorescences axillary, amongst the leaves or more usually below them on older wood, fasciculate with numerous single flowers in bud, but only 1–3 flowers open at one time; bracts 3–6, ± elliptic, 1–2 mm long, entire or bilobed, sometimes amplexicaul, pubescent; pedicel 4–6 mm long, 0.3–0.7 mm wide, with short greyish stellate hairs; articulation basal, inconspicuous. Flowers cream or yellow, fragrant, perianth divided for $^{1}/_{5}$ into 5(–6) ± ascending lobes, each (3–)3.5–4.5 mm long, 1.5–2(–2.5) mm wide, outer surface as flower stalk, inner minutely papillate. Male flowers with androphore terete, 1.8–2.5 mm, glabrous; anthers uniseriate, 8–10, glabrous, in a disc 0.5–1 mm long, 1–1.8 mm diameter, ovary vestigial, concealed. Female flowers with anthers barely reduced at the base of the ovary; ovary ± sessile, subglobose, ± 1 mm long, 1.5 mm diameter, densely tomentose; style ± 0.5 mm long; stigmas 4, recurved. Fruits red brown, borne in dense clusters, carpels one per fruit, ascending, longitudinally ellipsoid, 2–2.3 cm long, ± 1 cm wide, apex rounded, stipe 5 mm long, 3 mm wide; fruiting pedicel ± 4 mm long, glabrous.

TANZANIA. Rufiji District: Lake Utanga, fl. 15 Dec. 1971, *Ludanga* 1357!; Masasi District: forest above Ndanda mission, fl. 17 March 1991, *Bidgood, Abdallah & Vollesen* 2037!; Kilwa District: Kingupira, fl. 12 Dec. 1976, *Vollesen* in MRC 4216!
DISTR. **T** ?6, 6/8, 8; Mozambique ?
HAB. Woodland or thicket near lakes or rivers, rarely in forest; 100–500 m

LOCAL USES. None recorded.
CONSERVATION NOTES. Known to me from only seven fertile specimens. Threats to forest at some of its sites are documented in Clarke, Status Reports for 6 Coastal Forests in Lindi Region, Tanzania (1995). In view of these, an estimated area of occupancy of less than 2,000 km², the species this is here assessed as VU 2Ba, b(iii), i.e."vulnerable", in view of its scarcity and restricted distribution.
NOTES. The original authors of *Cola discoglypremnophylla* drew attention to the close relationship with "*C. clavata*" (i.e. *C. pseudoclavata*). They distinguished *Cola discoglypremnophylla* principally on the leaves being widely rounded and cordate or subcordate at the base, as seen in the holotype (see Brenan & Jones 1946, Fig. 1). However, the type specimen is aberrant, based on the fertile specimens that I have seen (Luke however disagrees) and this leaf shape is not typical of the species. Brenan was unaware of this aberration, rejecting a later collection by Eggeling of what appears to be this species, on the basis of leaf-shape. In fact, a normal specimen of *C. discoglypremnophylla* has leaves indistinguishable from *C. pseudoclavata* and the two can only reliably be distinguished in flower. The four principal discriminating characters are:

Pedicel 0.4–0.6 mm diameter, perianth lobes 5, ascending,
  androphore (2.5–)3–4 mm long . . . . *C. discoglypremnophylla*
Pedicel 0.2–0.3 mm diameter, perianth lobes 6–8(–10), ± reflexed,
  androphore 0.2 mm long . . . . *C. pseudoclavata*

Although no fertile collection has been seen by me from Mozambique (though Wild, *loc. cit.* refers to a specimen in bud and Luke (pers. comm. 2006) cites *Luke & Kibure* 10107 from the Mueda Plateau), it is very likely that a plant so widespread in the Lindi area should also occur over the border.
    Luke (pers. comm.) cites *Luke, Luke et al.* 7621 (EA n.v.), from Selous GR as firm evidence of the occurrence in **T** 6.

3. OCTOLOBUS

Benth. & Hooker, Gen. Pl. I: 982 (1867); Hutchinson in K.B. 1937: 394 (1937).

Monoecious evergreen trees with stellate indumentum. Wood deep yellow-brown; twigs terete, inconspicuously lenticellate, with bud-scale scars congregated at intervals. Leaves papery, sometimes in intermittent clusters on branches, alternate, simple, entire, pinnately nerved, the tertiary veins reticulate, conspicuous and raised below; petiole swollen and kneed at base and apex, both long and short on same stem; stipules linear or filiform. Flowers brittle, single, or few-fascicled and then only one flower open at a time, borne amongst the leaves, below the leaves or ramiflorous, subtended by 4–10 spirally inserted cup-like or flattened, ovate to oblong bracts; pedicel inconspicuous; perianth actinomorphic, uniseriate, the exterior densely pubescent, proximal half cylindrical or cup-shaped, the distal half divided into eight membranously margined ovate-triangular lobes. Male flowers with anthers included in the perianth tube, androgynophore stout, bearing 12–20 elongated anthers on the periphery of a disc-like head; carpels vestigial. Female flowers with androgynophore short, bearing presumed non-functional anthers at the base, ovary a globose head of 8–30 spirally arranged pubescent, stipitate, ellipsoid-cylindrical carpels; carpels with 5–10 anatropous ovules arranged in two ranks on the ventral wall. Fruit with up to 30 carpels in a dense globose capitulum; carpels shortly cylindrical to ellipsoid or obovoid, pericarp indehiscent, leathery-fleshy, pubescent, containing 1–3 seeds. Seeds glossy, black, ellipsoid, usually faceted by mutual compression, testa thin and fleshy, hilum large and circular; cotyledons thick, subequal, endosperm absent, radicle proximal to hilum.

Three species in Guineo-Congolian Africa of which one extends to E Africa. *Octolobus* is closely related to *Cola*, indeed, they are vegetatively indistinguishable, but distinct in the cylindrical perianth tube (absent or saucer-like in *Cola*, the (7–)8-lobed perianth (versus 5(–6)-lobed) and the numerous, spirally arranged carpels (versus 4–6 in one whorl).

Octolobus spectabilis *Welw.* in Trans. Linn. Soc. 27: 18 (1869); Germain in F.C.B. 10: 260 (1963); Cheek & Frimodt-Moeller in K.B. 53: 682 (1998). Type: Angola, Pungo Andongo, Barranca da Pedro Songue, *Welwitsch* 1202 (BM, holo.)

Tree 7(–15) m tall, ± 30(–60?) cm diameter at breast height; bole cylindrical, lacking buttresses, bark brownish with longitudinal fissures; stems of flowering branches terete or slightly ridged, 2–3.5 mm diameter, white to brown, black scurfy with stellate hairs, glabrescent, internodes 0.5–3(–5) cm long. Leaves elliptic, obovate or oblanceolate, 7–20(–30) cm long, 2.8–7.5(–13) cm wide, apex short to long-acuminate, acumen 10–15(–28) mm long, 4–5 mm wide, base attenuate and abruptly rounded-truncate or obtuse, lateral nerves 4–9 on each side of the midrib; petiole 0.5–5(–12) cm long; stipules 5–7 mm long, 1 mm wide. Inflorescence with bracts papery, broadly ovate to oblong, 1.5–8 mm long, 1–5 mm wide, apex truncate to mucronate, glabrous or densely velutinous. Flowers creamy-yellow, perianth outer surface densely golden-brown stellate-pubescent, with proximal tube short- or long-cylindrical, 4–11 mm long, 5–8(–20) mm wide, inner surface with thinly scattered patent, simple hairs; lobes 12–14(–22) mm long, 3–5(–10) mm wide, inner surface densely papillose, with a membranous margin ± 2 mm wide, glabrous. Female flowers with androgynophore 0.5–1 mm long, glabrous, carpels 8–30 in a globose mass ± 1 cm long, carpels oblong, 4 mm long, 1.5 mm wide, densely white tomentose, stigmatic heads glossy, black. Male flowers with androgynophore 3–5 mm long, 0.7 mm wide, glabrous, staminal head shortly cylindrical, 3 mm long, 4 mm wide. Fruit red, densely but shortly tomentellous, with accrescent pedicel and androgynophore combined 1.5 cm long, stout; carpels obovoid or shortly ellipsoid-cylindrical, 1.4–3.5 cm long, 1.1–2.8 cm wide, with a wide raised ventral ridge, stipe (0.2–)0.8–1.5 cm long, 0.2–0.3 mm wide, rostrum 0.4–0.8 cm long, 0.2–0.3 cm wide. Seed glossy black, ellipsoid, often faceted, 1–2.1 cm long, 0.8–1.2 cm wide, hilum ± 1 cm diameter. Fig. 6, p. 47.

H.Fitch, del et lith.

J.N Fitch, imp.

FIG. 6. *OCTOLOBUS SPECTABILIS* — **1**, habit × 1; **2**, female flower, opened out × 1; **3**, gynoecium
of female flower, longitudinal section, × ⁶/₅; **4**, anther from female flower, side view × 10; **5**,
anther from female flower, front view × 10; **6**, carpel from female flower, longitudinal section
× 5; **7**, male flower, opened out × 1; **8**, mericarp, from fruit, side view, × 1; **9**, mericarp, with
seeds exposed, × 1; **10**, seed, showing hilum, × 1; **11**, seed, longitudinal section, × 1; **12**, embryo
of seed with one cotyledon detached, × 1; **13** radicle and plumule of embryo × 3. All drawn
from *Welwitsch* 1202. Reproduced from T.L.S. 27: t.VI. Drawn by W.H. Fitch.

Tanzania. Iringa District: Udzungwa Scarp Forest Reserve, 35° 58' E, 8° 21' S, Dec. 1997, *Frimodt-Moeller, Ndangalasi & Joeker* TZ 540!
Distr. **T** 7; Sierra Leone to W Congo-Kinshasa and Angola
Hab. Forest with *Allanblackia* and *Afrothismia*; ± 950 m

Syn. *O. angustatus* Hutch. in K.B. 1937: 395 (1937). Type: Ghana, Kwahu Prasu, *Vigne* 1602 (K!, holo.)

Conservation notes. This species though never very common, is widespread in Africa and is known from more than 15 sites. While there are some threats to its habitat at the single E African site, threats at other sites have not been evaluated. It is rated here as "Near Threatened".
Notes. This genus is so far represented in E Africa by a single specimen, overgrowing a stream in a steep, inaccessible gorge where several other new or rare species have been found, for example *Afrothismia sp.*, *Ancistrocladus tanzaniensis* Cheek & Frimodt-Moeller, *Trichilia lovettii* Cheek and *Cola stelechantha* Brenan. This collection represents an extreme variant of the polymorphic *Octolobus spectabilis*, having much larger leaves (to over 30 × 11 cm) than any specimen known from W Africa where leaves are usually no more than ± 20 × 8 cm. Furthermore, the leaves of the Tanzanian collection are elliptic, with a rounded to obtuse base, rather than being obovate, or narrowly elliptic with an acute or cuneate base. However, since this species is already known to be exceedingly variable in perianth shape and indumentum, and in stipule shape, size and persistence, it would be unsatisfactory to recognize this particular variant without recognizing the others that occur.
 This species was recently reported by Mbago (pers. comm. to Luke 2006) from West Kilombero scarp, Forest Reserve Nyumbanitu-Udekwa, fr. Dec. 1999, 1000m alt. (*Mhoro and Frontier in Mbago* 1822) but I have not seen the specimen. The species is also listed in the Mtwara report of Frontier Tanzania, but there is no specimen to substantiate this claim, so the record must be treated as dubious.

4. **PTERYGOTA**

Schott & Endl. in Melet. Bot. 32 (1832)

Monoecious, deciduous or evergreen trees, with stellate indumentum. Wood white, soft and light. Leaves alternate, simple, entire; domatial pockets sometimes present; petiole usually swollen and kneed at base and apex; stipules early caducous. Inflorescences several per stem in the uppermost axils, bearing short partial peduncles, partial peduncles 1-several flowered; bracts caducous; pedicels articulated with the partial peduncle. Flowers actinomorphic, brown and purple, perianth of a single whorl of five perianth lobes united at base; androgynophore terete. Male flowers with 5–20 anthers arranged in a dense head of one or two whorls or spirally arranged, concealing vestigial carpels. Female flowers fewer, usually at the ends of the inflorescence, androgynophore abbreviated, anthers often as numerous and as large as the male, but not dehiscing; ovary of five coherent, ovoid carpels each with a filamentous style and truncate stigma, carpels each with numerous ovules in two superposed ranks. Fruit with the carpels strongly apocarpous and stipitate; carpels large, thickly woody, dehiscent along a ventral suture revealing 20–30 seeds arranged in two files, one on each side of the suture. Seeds dry, large, the distal end produced into a papery or spongy wing.

Pantropical, 15–20 species.

Pterygota alata (*Roxb.*) *R.Br.*, Pl. Jav. Rar. 234 (1844). Syn.: *Sterculia alata* Roxb., Hort. Bengal. 50 (1814), *nom. nud.*; Roxb., Pl. Coromandel 3: 84, t. 287 (1820); U.O.P.Z.: 455 (1949)
 Native to India. Cultivated in Victoria Gardens, Zanzibar.

Leaves broadly ovate, shallowly 3–5-lobed, 12–30 × 10–23 cm,
　　lower surface with conspicuous domatia; **U** 1, 2, 4; **T** 1, 4;
　　750–1550 m .... 1. *P. mildbraedii*
Leaves lanceolate, entire or obscurely lobed, 13–19 × 8–13 cm,
　　lower surface lacking domatia; **T** 3, 8; 0–400 m .... 2. *P. perrieri*

1. **Pterygota mildbraedii** *Engl.* in Z.A.E.: 506 (1912); Wild in F.Z. 1: 561 (1961); Germain in F.C.B. 10: 261, f. 23 (1963); Troupin, Fl. Pl. Lig. Rwanda: 666, f. 228.2 (1982); Hamilton, Uganda For. Trees: 116 (1981); Troupin, Fl. Rwanda 2: 406, f. 127.2 (1983). Type: Rwanda, near Lac Muhazi, *Mildbraed* 581 (B, presumed destroyed, holo.)

Deciduous tree 30–40 m tall; trunk buttressed, grey, smooth; flowering branchlets ± 5 mm diameter, glabrous, lacking conspicuous lenticels. Leaf-blade broadly ovate in outline, shallowly (3–)5-lobed, 12–30 cm long, 10–23 cm wide (smaller leaves from flowering branchlets), apex shortly acuminate, base cordate to deeply cordate, digitately 5-nerved, glabrous above, and below when mature (expanding leaves and stems densely and entirely covered with 10–12-armed golden brown stellate hairs, these persisting only on the apex of the petiole), domatial pockets webbed, stellately hairy, conspicuous; petiole (6–)12–17 cm long, (0.1–)0.2–3 mm diameter Inflorescences 2–5 per branchlet, golden brown stellately tomentose as the young leaves, borne singly in the uppermost axils, each (2.7–)5–6 cm long, 2 mm diameter, bearing 3–5 partial peduncles; bracts orbicular, concave, 9 mm long, partial peduncle 8–12 mm long, bearing 2–3 bracteoles similar to the bracts and at the apex ± 3 flowers, pedicel ± 4 mm long. Male flower with outer indumentum as inflorescence, divided for $^5/_6$, lobes oblong, 10–11 mm long, 4 mm wide, basal part saucer-shaped, 2 mm long, 5 mm wide, inner surface densely papillate, upper half of lobes densely stellate; androgynophore 5 mm long, glabrous, base dilated, surrounded by stellate hairs, apex with subcylindrical head of anthers 2.5 mm long, 2.5 mm wide. Female flowers as the male, androgynophore <1 mm, carpels, stamens forming a cylinder ± 4 mm diameter; carpels ovoid, ± 6 mm long, 3–4 mm wide, villose, styles slender, ± 4 mm long, revolute. Fruit with carpels ellipsoid, 9–11 cm long, 8–12 cm wide, carpel wall 9–17 mm thick, outer surface indumentum as inflorescence; seeds with wing obovate to oblong, 8.5–10 cm long, 2.8–3.5 cm wide.

UGANDA. West Nile District: Zeu [Zeio], March 1935, *Eggeling* 1913!; Ankole District: Kalambalu, 19 June 1905, *Dawe* 441!; Mubende District: Mubende Hill, Feb. 1933, *Brasnett* 1316!
TANZANIA. Biharamulo District: Katoke Mission, 4 June 1956, *Gane* 73! & Biharmulo–Kibondo road, 9 km, July 1951, *Eggeling* 6227!; Kigoma District: Mt Livandabe, 28 May 1997, *Bidgood et al.* 4160!
DISTR. **U** 1, 2, 4; **T** 1, 4; Cameroon, Congo-Kinshasa, Rwanda, Burundi, Zambia
HAB. Evergreen forest; 750–1550 m

SYN. ?*Pterygota alata* sensu K.Schum., E.M. 5: 135 (1900), *non* (Roxb.) R.Br.
　　Pterygota schumanniana Engl., V.E. 3(2): 469 (1921); T.T.C.L.: 601 (1949). Type: Tanzania, probably Tabora District: Kavendo, Ratuma R., *Bohm & Reichard* 63 (B, presumed destroyed, holo.), **syn. nov.**
　　Pterygota macrocarpa sensu Robyns, F. P. N. A.: 612, f. 29 (1948), *non* K.Schum.
　　Pterygota sp. nov. of I.T.U.: 422 (1952)

CONSERVATION NOTES. This species appears to be widespread and common, and its habitat not significantly threatened as far as is known. It is provisionally rated here as of "least concern" for conservation.
NOTES. The identity of Schumann's *?Pterygota alata* and of *Pterygota schumanniana* Engl., which was based upon it and validated by Schumann's description (Art. 33 of the Code, see Note 2 and Example 9) is almost certainly *P. mildbraedii*. This conclusion is contrary to that reached by Wild (F.Z. 1: 562, 1961) who attributed these names to the lowland coastal species. The

main evidence against Wild is that Schumann (*loc. cit.*) cites a specimen from a locality that is clearly in western Tanzania, near the inland lakes and amongst the mountains and is thus within the distribution of *P. mildbraedii*, the only species of the genus to occur there. I have not traced the specimen, nor the exact locality names (Kavendo, Ratuma R.) cited by Schumann. However, it is likely that Ratuma R. is a misspelling for Katuma River, which occurs in **T** 4.

2. **Pterygota perrieri** *Hochr.* in Candollea 3: 148 (1926); Arènes in Fl. Madag. 131: 18, t. 6 (1959). Type: Madagascar, Iabohazo R. banks, *Perrier* 1395 (P, holo.; K!, iso.)

Tree 18–36 m tall, buttressed to 4.5 m high; flowering branchlets 3–4 mm diameter, glabrous when mature. Leaf-blade of flowering branchlets digitately 3–5-nerved, lanceolate, entire or obscurely 5-lobed, rarely shortly oblong, shallowly 5-lobed, 13.5–19 cm long, 8.5–13 cm wide, apex acute to rounded, base truncate or slightly cordate (on vegetative shoots larger, ovate, to 26 × 21 cm, base cordate, petiole 6–7.5 cm long, 3 mm diameter); both surfaces with inconspicuous, thinly scattered stellate hairs, domatia absent; petiole 4.5–6.5 cm long, 1–1.5 mm wide, glabrous with indumentum as blade when young, glabrescent. Inflorescences axillary in the topmost nodes, indumentum denser than leaf-blade, puberulous, 2–4(–6) inflorescences per branch, each (2.5–)4–4.5 cm long, bearing ± 5 partial peduncles; bracts highly caducous, not seen; partial peduncle 1(–8) mm long, 1(–2)-flowered; pedicel 2 mm long. Male flowers with outer indumentum as inflorescence, divided for $^9/_{10}$, lobes oblong, 16–19 mm long, 2.5–3.5 mm wide, basal part saucer-shaped, 2 mm long, 5 mm wide, inner surface densely invested with a mixture of papillae and stellate hairs; androgynophore 7–9 mm long, lower third minutely puberulous, upper third glabrous, 0.5 mm diameter, base dilated, conical, apex with subcylindrical head of anthers 2.5 mm long, 2.5–4.5 mm wide. Female flowers as the male, androgynophore <1 mm, stamens forming a cylinder 4 mm diameter; carpels ovoid, 5 mm long, 2–2.5 mm wide, puberulous, styles slender, 4.5 mm long, revolute. Fruit with carpels ellipsoid, 7.5–8(–13) cm long, 7–8 cm wide, carpel wall 5–6 mm thick, outer surface indumentum as inflorescence; seeds with embryo area highly wrinkled, wing deep brown, obovate to oblong, 6.5–7 cm long, 2.5–4.2 cm wide. Fig. 7, p. 51.

TANZANIA. Pare District: Ganja Maore–Same, July 1955, *Semsei* 2130!; Lushoto District: Mombo, 9 Nov. 1955, *Milne-Redhead & Taylor* 7264!; Kilwa District: Mto Nyangi Hippo pools, 28 Nov. 2003, *Luke & Kibure* 9760!
DISTR. **T** 3, 8, ?Zanzibar; Mozambique, Madagascar, Comores (Mayotte)
HAB. Riparian forest, sometimes on limestone; 0–400 m

SYN. ?*Pterygota alata* sensu Wild in F.Z. 1: 562 (1961), *non* K.Schum.
     P. schumanniana sensu Wild in F.Z. 1: 562 (1961), *non* Engl.

CONSERVATION NOTES. Six specimens are cited in the Flore de Madagascar, all from locations in the western half of the island. Despite numerous collection trips to Madagascar in the last 20 years by several institutes, no new specimens have come to light. In 2001 the taxon was discovered in Mayotte (*Pascal* 717). In E Africa, ± 10 sites are known from almost as many specimens. The single specimen seen from Zanzibar (*Robins* 93, K!) is sterile, therefore a degree of uncertainty attaches to its identification. The stronghold for this species appears to be **T** 3 where eight specimens from five districts are known. Where noted, specimens represent either single trees or a small grove of up to five individuals. *Pterygota perrieri* appears to be a timber species. This may constitute a threat to its survival. Threats at one site, Kwamgumi, have been recorded (Kew Bull. 57: 417, 2002) but are unknown at its other sites. On the basis of the data above, the taxon is here assessed as Near Threatened. Survey work is recommended to assess the conservation status of this species properly.
NOTES. The E African material has long been regarded as a possible new species but on the evidence available to me I am unable to distinguish it from *P. perrieri*, previously believed to be endemic to Madagascar. While the material in Madagascar tends to have more slender, 3-nerved leaves than the E African material (usually 5-nerved), this is not clear cut. I have found no distinguishing characters in the flowers. Unfortunately I have not had access to

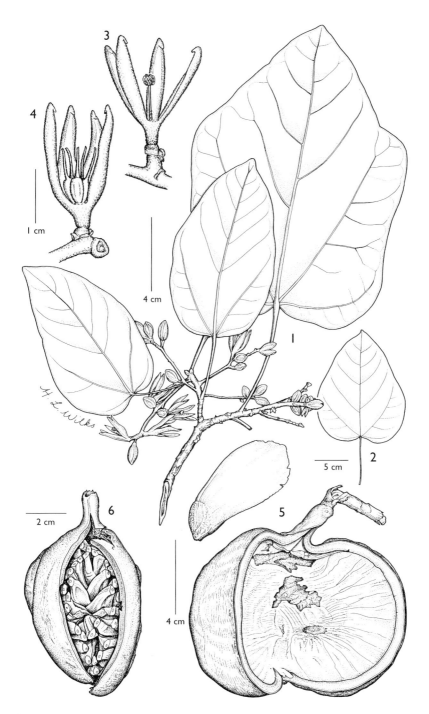

Fig. 7. *PTERYGOTA PERRIERI* — **1**, habit, flowering branch; **2**, smaller leaf; **3**, male flower, side view, one tepal removed; **4**, female flower, one tepal removed; **5**, mature dehisced mericarp, with separate seed (above); **6**, immature fruit, precociously dehisced, showing seeds. 1, 3 & 4 from *Eggeling* 6692; 2 & 6 from *Luke et al.* 7504; 5 from *Milne-Redhead & Taylor* 7264. Drawn by Hazel Wilks.

fruiting material from Madagascar. Since fruit characters can be important at the species level in the genus, it may be necessary to reassess the taxonomic status of the E African material should these become accessible. The habitat of the E African and Malagasy material seems to be identical.

5. HERITIERA

Dryand. in Aiton, Hort. Kew., ed. 1, 3: 546 (1789); Kosterm. in Madj. Ilmu Pengetah. Indonesia 1: 1–121 (1959) & in Reinwardtia 4(4): 465–583 (1959); Bayer & Kubitzki in Fam. Gen. Vasc. Pl. 5: 265–266 (2003)

Trees with well-developed buttresses; boles erect, often much branched; young stems and leaves below usually covered with small, fimbriate, peltate appressed scales. Leaves palmately compound and 1–6-foliolate or simple (FTEA); palmately 3(–5)-nerved from the base, pinnately nerved above; petiole often pseudopeltate (leaf base attached above the petiole); stipulate. Flowers unisexual in many-flowered, monoecious, axillary panicles; peduncle and main ramifications lepidote, the tomentum gradually changing to stellate hairs toward the ultimate ramifications; pedicels articulate; bracts and bracteoles present. Calyx campanulate (FTEA) or urceolate, 4–5(–6)-lobed. Petals 0. Male flowers: androgynophore present with 4–5 anthers in a ring (or irregular clump) at its apex, anthers di-thecal, extrorse, longitudinally dehiscent, sometimes minute sterile carpels protruding through the ring of anthers. Female flowers: perianth as in the male, but slightly larger; androgynophore absent or short; ovary of 4–5(–6) sessile, glabrous (FTEA) or lepidote, connate carpels; base of the ovary surrounded by sterile anthers equal in number to the carpels; styles connate, short, spreading or incurved; stigmas 5, minute or ± thickened; 2 ovules per locule. Fruit with 1–4 carpels each maturing into a 1-seeded woody to spongy fibrous indehiscent samara, the wing or ridge apical, ± well-developed or weakly-developed (FTEA). Seeds glabrous; exalbuminous; cotyledons 2, small radicle next to hilum and plumule turned to the ventral side.

A genus of 30–35 species, principally tropical and subtropical Asia, but also Australia, Pacific Islands, and tropical Africa (3 species, 2 on the west coast and 1 on the east coast).

Heritiera littoralis *Dryand.* in Aiton, Hort. Kew., ed. 1, 3: 546 (1789); Mast. in F.T.A. 1: 225 (1868); K. Schum. in Engl., P.O.A. C: 272 (1895); K. Schum. in E.M. 5: 136–137, t. 10, fig. C/a-f (1900); Engl., V.E. 3(2): 470, fig. 217 (1921); T.S.K.: 42 (1936); T.T.C.L.: 599 (1949); U.O.P.Z.: 294 (1949); Wild in F.Z. 1(2): 564, t. 106 (1961); K.T.S.: 548–549 (1961); Wild & Gonç. in F.M. 27: 55–57, t. 7 (1979); K.T.S.L.: 165, fig., map (1994). Types: Sri Lanka [Zeylon], *Koenig* s.n. (BM, syn.); Vietnam [Pulo Condore], *Nelson* s.n. (BM, syn.); and Rheede, Hort. Malab. 6: 37, t. 21 (1686) (syn.). One of the original syntypes, "Samandura. *Linn. zeyl.* 433", is excluded since it represents *Samadera indica* Gaertn. (Simaroubaceae).

Large, evergreen trees, (3–)10–25 m tall, to 1 m DBH; low-branched, crowns dense; base of bole with long and sinuous plank buttresses, to 30 cm tall; bark pale grey, furrowed or reticulate, slash pink; young branchlets lepidote, but soon glabrescent. Leaves elliptic to ovate or elliptic-oblong, 5–22(–30) cm long, 2.5–11(–24) cm wide, apex acute, subobtuse or obtuse, margin entire or undulate, base rounded to subcordate, often oblique; silvery-grey below, covered with a dense layer of minute, fimbriate, appressed scales, the centres of which are brown, occasionally completely brown scales scattered among the silvery scales, green and glabrous above; palmately 3(–5)-nerved from the base, midrib and 2° veins prominent below, other veins obscure; midrib slightly raised above, 2° veins slightly impressed; coriaceous; petiole angular, stout, ± pulvinate, 0.8–2.1 cm long, lepidote; stipules lanceolate-subulate, ± 3 mm long, lepidote, caducous. Inflorescences to

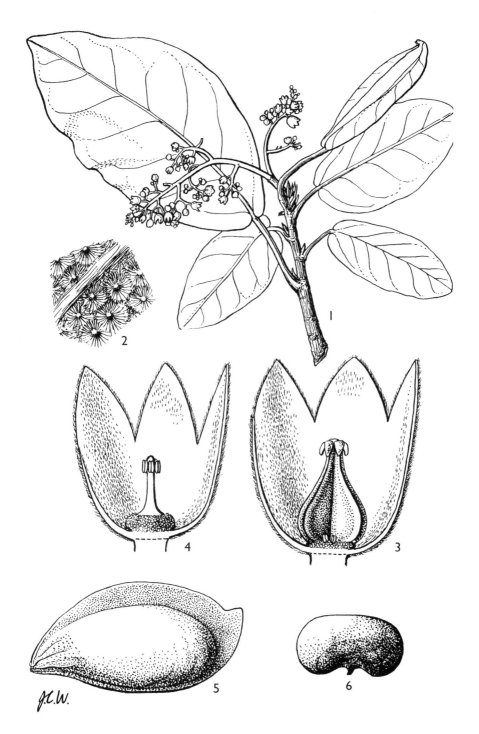

FIG. 8. *HERITIERA LITTORALIS* — **1**, habit × ²⁄₃; **2**, lower surface of leaf × 24; **3**, female flower in section × 10; **4**, male flower, in section × 10; **5**, fruit, side view × ²⁄₃; **6**, seed × ²⁄₃. 1–4 from *Kirk* s.n.; 5 & 6 from *Faulkner* 750. Reproduced from F.Z. 1: t. 106. Drawn by Joanna Webb.

11(–18) cm long; the base of the main peduncle and lower ramifications lepidote, the remainder densely stellate-pubescent; pedicels to ± 5 mm long, articulated below the calyx; bracts and bracteoles to 4 mm long, ovate, acute, tomentose. Floral buds globose. Flowers yellowish-green or greenish-brown, aromatic. Male flowers: calyx campanulate, 3–6 mm long, divided for ± $^1/_3$, lobes 4(–5), ovate-lanceolate, acute, outer surface densely stellate-pilose, inner surface pilose at the base, but becoming stellate-pubescent toward the mouth; androgynophore ± 1 mm long, dilated at the base and surrounded by a minute, glandular cushion-like disc, with a ring of 4–5 anthers in a single series below the apex and a vestigial style exserted from the centre of the ring. Female flowers: as in the male, but slightly larger, 3–7 mm long; carpels 4–5, connate into a widely ovoid, glabrous ovary that tapers gradually to the styles, and which is surrounded at the base by (4–)5 free, rudimentary sessile stamens; styles ± 0.5–1 mm long, connate; stigmas recurved. Fruiting carpels free, indehiscent, 1–4(–5) maturing, oblong-ovoid, 3.7–9 cm long, 3–6 cm in diameter, 3.5–5.5 cm tall (including ridge or wing), woody, glabrous, glossy light brown, flat ventrally, strongly ridged dorsally, the wing to 4–10 mm long, the inner fruit wall densely appressed stellate-pilose; 1 seed per carpel; seed flattened towards the ventral suture, oblong-ellipsoid, ± 2.5–3 × 2.5–3.5 × 1.2–2 cm, testa light brown, rugose (an artifact of drying?), glabrous; cotyledons 2, thick and fleshy. Fig. 8, p. 53.

KENYA. Tana River District: Ribe, 22 Aug. 1936, *Abdulla* 1139!; Mombasa, Port Tudor, Nov. 1931, *MacNaughton* 9246!; Kwale District: S of Kinondo, 2 Mar. 1977, *Faden & Faden* 77/646!

TANZANIA. Tanga District: Tanga Bay, 19 Dec. 1946, *Greenway* 7897! & idem, 14 Nov. 1947, *Brenan & Greenway* 8323!; Pangani District: Bushiri Estate, 14 Nov. 1950, *Faulkner* 750!; Pemba: Tundaua, June 1928, *Vaughan* 350!

DISTR. **K** 7; **T** 3, 6, 8; **P**, **Z**; Mozambique; Comoro Islands, Madagascar, Seychelles, and introduced into the Mascarene Islands; also widely distributed along the coasts of tropical Asia, Australia, and Pacific Islands

HAB. Mangrove swamps, landward side where fresh water intermingles with the salt water of the sea, locally common or dominant; also on sand at the tide line and in forest on coral rag; 0–5 m

SYN. *Heritiera littoralis* Dryand. subsp. *ralima* Arènes in Mém. Inst. Sci. Madag., sér. B, 7: 78 (1956) & in Fl. Madag. 131: 34, fig. 9/6–7 (1959). Type: Madagascar, Lavatsiraka-Sahambavany, *Service Forestier* 2769-SF (P, holo.; K!, iso.)

LOCAL USES. Trunks of this species are used for dhow masts in Lamu (*Abdulla* 1139) and Zanzibar (U.O.P.Z.) and were formerly used for railroad sleepers (*Farquhar* 6).

CONSERVATION NOTES. This species is widespread and common in mangrove, and its habitat not significantly threatened as far as is known. It is provisionally rated here as of "least concern" for conservation.

NOTES. The capsules are common in drift along the shore. The dorsal ridge of the capsule, often described as a keel, is homologous with a wing and actually functions as a sail. Fruits float on the surface of the ocean always with the ridge or wing upward and they are widely distributed by a combination of currents and wind.

　　　The record from **T** 8 is based on a sight record by Luke (pers. comm.) at Mtwara Mnazi Bay/Ruvuma Est MNP).

6. **HILDEGARDIA**

Schott & Endl., Melet. Bot. 33 (1832); Dorr & L.C. Barnett in K.B. 45: 577–580 (1990); Cheek & G. J. Leach in Kew Mag. 11: 88–94 (1994)

Deciduous shrubs or trees, often with swollen trunks. Leaves entire or slightly 3-lobed; palmately nerved from the base; domatia present or not; petiolate; stipulate. Inflorescences axillary or subterminal, flowering either when leafless (FTEA) or when ± leafy. Flowers bisexual or functionally unisexual with vestigial anthers. Calyx gamosepalous, elongate-campanulate or fusiform-clavate, 4–5-lobed, the lobes short or long, often reflexed, sometimes persistent in fruit. Petals 0. Stamens fused in a

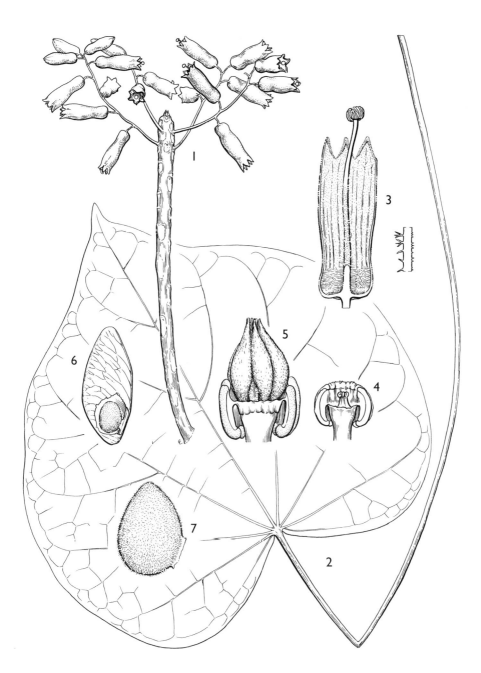

FIG. 9. *HILDEGARDIA MIGEODII* — **1**, habit, leafless branch with inflorescences, × ²/₃; **2**, leaf × ²/₃; **3**, male flower, sagittal section, with detailed section of perianth wall, × ²/₃; **4**, detail of male flower, showing anthers and vestigial carpels, × ²/₃; **5**, detail of female flower, showing carpels and fully formed anthers, × ²/₃; **6**, mericarp, with window to show the seed, × ²/₃; **7**, seed, side view, × 2. 1, 2–7, from *Migeod* 562; 2 from *Migeod* 662. Drawn by Eleanor Catherine.

globose head borne on a slender androgynophore; anthers 10–20, linear, sessile or subsessile, di-thecal, surrounding the carpels or pistillode. Ovary of 5 coherent carpels (becoming free in fruit); 1–2 ovules per carpel; styles coherent, sometimes very short; stigma lobes 5, reflexed. Fruiting carpels usually 5, stipitate, winged, somewhat inflated, membranaceous, dry, indehiscent, the seed(s) basal. Seeds 1(–2), glabrous or pubescent; albuminous; radicle next to hilum.

Eleven species with a remarkable pantropical distribution; one from Cuba, three each from Africa and Madagascar, and one each from India, Indonesia (Lesser Sunda Islands), the Philippines (Mindoro Island), and Australia (Northern Territory).

The African and Malagasy species appear to form a natural group. All have long, tubular or infundibuliform calyces with relatively short calyx lobes, and they are leafless when in flower. Molecular data (Wilkie et al. in Syst. Bot. 31: 160 (2006)) fail to support the current generic circumscription of *Hildegardia*; limited molecular sampling supports our argument based on floral morphology that the African and Malagasy species of *Hildegardia* form a natural group.

Hildegardia migeodii (*Exell*) *Kosterm.* in B.J.B.B. 24: 337 (1954); Wild in F.Z. 1(2): 562 (1961); Wild & Gonç. in Fl. Moçamb. 27: 54 (1979). Type: Tanzania, Lindi District: Tendaguru, *Migeod* 562 (BM!, holo.)

Deciduous trees, to 12–15 m tall; bark greyish; glabrous except for the interior of the domatia and calyx, and the androgynophore and seeds. Leaves cordate to very widely cordate-ovate, 13–19 cm long, 12.5–18.5 cm wide, apex acuminate (obtuse fide F.Z.), margin entire and revolute, base cordate to widely cordate, glabrous above and below but with pocket-like domatia in the axils of the 1°, 2°, and 3° veins below, the interior of the domatia with simple hairs; palmately 7–9-nerved from the base; petioles terete, 14–32 cm long. Inflorescences axillary racemes (as evidenced by petiole scars), 4.5–7 cm long, borne near the ends of branches; pedicels 3–4 mm long, articulated near the apex to form a 2 mm long stipe at the base of the calyx; bracts fugacious, not seen. Flowering when leafless. Calyx elongate-campanulate, slightly curved, 2–2.2 cm long, 0.6–0.8 cm wide, dilated below, base truncate, subtended by a 1–2 mm long stipe, orange to red or salmon-coloured, glabrous externally, villous at the base internally, becoming finely pubescent above and tomentellous along the margins of the calyx lobes, lobes acute, 3–4 mm long, 2–2.5 mm wide, erect or weakly reflexed. Androgynophore 1.5–2.5 cm long, villous at the base, finely pubescent above and glabrous on the exserted portion; exserted 5–10 mm beyond the calyx lobes; anthers 15, linear, in 2 rows congested in a globose head. Ovary of 5 coherent carpels, ovoid; 1–2 ovules per carpel; stigma lobes recurved, subsessile. Fruit apocarpous, carpels borne on the elongated androgynophore, elliptic to lanceolate, (3.5–)5.2–5.5 cm long, (1.7–)2.1–2.3 cm wide, purplish, membranaceous, glabrous, inner walls of follicles sparsely pubescent with simple hairs near the attachment of the seed, the basal stipe ± 2 mm long, persisting; seeds 1(–2?), ovoid, ± 1 cm in diameter, hispid, with simple, hyaline hairs. Fig. 9, p. 55.

TANZANIA. Rufiji District: Selous Game Reserve, Stiegler's Gorge, fl&fr. Aug. 1993 *Luke & Luke* 3668!; Lindi District: Tendaguru, fr. 3 Aug. 1919, *Migeod* 585! & Lindi N, near Mchinga, ster. Dec. 2003, *Luke & Kibure* 10206!
DISTR. **T** 6, 8; Mozambique
HAB. Wooded grassland, in fringing bushland; 40–250 m

SYN. *Firmiana migeodii* Exell in J.B. 68: 83 (1930); Exell in J.B. 69: 100 (1931)
    Erythropsis migeodii (Exell) Ridl. in K.B. 1934: 216 (1934); T.T.C.L. 599 (1949)

CONSERVATION NOTES. Four sites only are known for this species, at one of which, at least, it is only known from a single tree (Luke pers. comm.). Trees in unprotected areas in coastal Tanzania are under threat of wood extraction. In view of this, *H. migeodii* is here assessed as Endangered (EN B2 a, b(iii)). The tree is genuinely rare. Botanists surveying in the vicinity of the type have not found the species (Vollesen pers. comm.)

Notes. F.Z. and Fl. Moçamb. describe the sole collection of *H. migeodii* from Mozambique as a shrub, ± 2 m tall.

H. gillettii Dorr & L.C.Barnett, a tree, is to be expected in northern Kenya since its type and the few other collections were made nearby in Somalia. It can be distinguished readily from *H. migeodii* by its shorter (3.4–5.2 cm long), stellate-pubescent petiole; non-revolute leaf margins; obtuse to acute leaf apex; stellate-pubescent pedicels; and short (less than 1 mm long), inconspicuous stipe subtending the calyx.

7. MANSONIA

J.R.Drumm. in J.L.S. 37: 260 (1905); Prain in J.L.S. 37: 250–263, t. 10 (1905); Bayer & Kubitzki in Fam. Gen. Vasc. Pls. 5: 261 (2003)

Large evergreen or deciduous (?) trees. Leaves entire, margin remotely crenate to faintly dentate, palmately nerved from the base, domatia present or not; petiolate; stipulate. Inflorescences axillary or subterminal, many-flowered, corymbose cymes, the ultimate divisions subumbellate; peduncle and pedicels expanding in fruit; subtended by inconspicuous, bracts and bracteoles caducous. Floral buds ovate-acute. Flowers mostly hermaphroditic, rarely male, zygomorphic. Calyx spathaceous, splitting unilaterally, deciduous. Petals 5, contorted-imbricate, opposite the carpels, free, glabrous above, ciliate at the base and contracted into a small claw or claw absent, white or yellow to reddish; small, scale-like nectary attached at the base of the inner surface of the petal (FTEA) or nectary absent. Androgynophore well-developed; stamens 10, uniseriate, paired or distinct, inserted at the apex of the androgynophore, filaments free or almost free, much longer than (FTEA) or shorter than the anthers; anthers mono- or di-thecal (FTEA), superposed when di-thecal, dorsifixed, extrorse; staminodes 5, petaloid, lanceolate to linear, valvate, borne in a whorl between the stamens and carpels and alternate with the carpels. Ovary of 5 free carpels, carpels fusiform, densely-pubescent, opposite the petals; 5–9 ovules per carpel, ovules biseriate, minute, anatropous, affixed to the inner suture of the carpel; each carpel prolonged into a filiform, flexuous style; stigmas minute, capitate. Fruit dry, 1–2(–3) carpels maturing into 1-seeded, indehiscent samaras, the wing dorsal, opposite the raphe and funiculus of the seed, chartaceous, pubescent internally. Seeds attached above, glabrous; endosperm scanty; cotyledons 2, thin, contortuplicate, but when teased apart large, obcordate, foliaceous, and prominently palmately-veined.

Four or five species with a remarkable distribution; one species (including two varieties) extends from Guinea–Conakry to Nigeria and Cameroon, one (doubtfully distinct from a variety of the former) is endemic to Cameroon, one is endemic to Tanzania, one is endemic to India (Assam) (but not treated by Malick in Fl. India 3: 407–476(1993)), and one is known from Myanmar and Thailand.

The East African species stands apart from the other species in the genus in having di-thecal (versus mono-thecal) anthers. This character is the principal basis for dividing the genus into *M.* subgen. *Diatomanthera* Brenan, which consists solely of *M. diatomanthera*, and *M.* subgen. *Mansonia* (*Eu-Mansonia* A.Chev.), which accomodates the remaining species.

ING, Tropicos, and Bayer & Kubitzki (2003) cite the authorship of *Mansonia* as "J.R.Drumm. ex Prain", but Prain is clear in his article on the Mansonieae that the author of the generic name is "J.R.Drumm." alone.

Mansonia diatomanthera *Brenan* in Hook. Ic. Pl., ser. 5, 5: 3451 (1947); T.T.C.L.: 601 (1949); Verdc. in K.B. 1953: 89, fig. 1 (1953); Kokwaro, Medic. Pl. E. Afr.: 209 (1976). Type: Tanzania, Handeni District: Kideleko, *Yussif bin Mohamedi* 8626 (K!, holo.; K!, iso.)

Large deciduous (?) trees, 15–20 m tall; young branchlets densely pubescent with dark brown stellate hairs, soon glabrescent; older branchlets knobby, glabrous; bark lead-gray, faintly longitudinally furrowed and conspicuously transversely cleft. Leaves

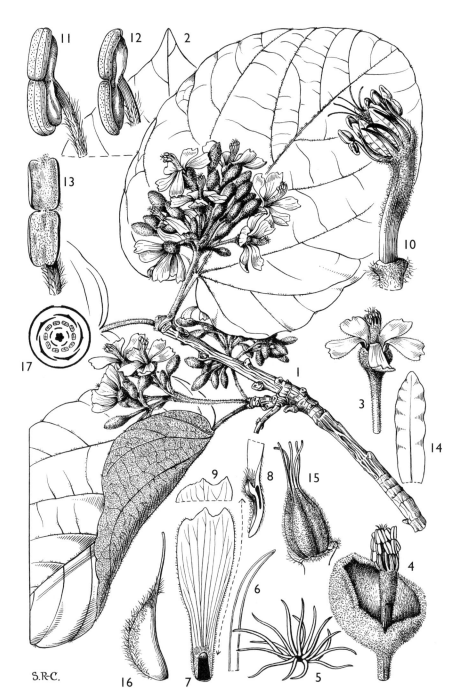

Fig. 10. *MANSONIA DIATOMANTHERA* — **1**, habit × 1; **2**, leaf tip × 1; **3**, flower × 1; **4**, calyx and androgynophore × 4; **5**, stellate hair from calyx × ± 60; **6**, arm from stellate hair × ± 120; **7**, petal, inner surface × 4; **8**, petal base in longitudinal section showing nectary × 8; **9**, petal apex × 4; **10**, androgynophore × 6; **11**, **12**, petal in three-quarter front view and side view, respectively × 12; **13**, anther, dehiscing × 12; **14**, staminode × 12; **15**, gynoecium × 8; **16**, carpel × 12; **17**, floral diagram. All from *Yussif bin Mohamedi* 8626. Reproduced from Hook. Ic. Ser. 5, 5: t. 3451. Drawn by Stella Ross-Craig.

crowded at the apices of the principal branches and on short, lateral shoots, broadly ovate to ovate-elliptic or subovate, 7.5–20 cm long, 6–12.5(–17.5) cm wide, apex acuminate, cuspidate or rounded, margin remotely crenate, base cordate, oblique, the sinus widely open, coriaceous (chartaceous when young), densely stellate-pubescent below, sparsely pubescent with stellate, fasciculate, or rarely simple hairs confined to the midrib and veins above; palmately 7–10-veined from the base, lateral veins ending in rounded teeth; petiole compressed, 1–2(–3) cm long, stellate-pubescent, soon glabrescent; stipules subulate, ± 5 mm long, densely stellate-pubescent, caducous. Inflorescences axillary, corymbose cymes, the ultimate divisions umbellate; peduncle 0.8–2 cm long, expanding in fruit, densely stellate-pubescent, soon glabrescent; pedicels ± 5–10 mm long, indument as peduncle; bracts and bracteoles stellate-pubescent, caducous. Calyx spathaceous, ± 8–9 mm long, apex subacute to abruptly narrowed, outer surface densely ferrugineous stellate-pubescent, splitting unilaterally just before anthesis. Petals yellow, oblong-oblanceolate, 10–13 mm long, 3.5–4 mm wide. Carpels fusiform, ± 2 mm long, ± 0.75 mm wide, densely tomentose with easily removed stellate hairs; style filiform, flexuous, 2–2.5 mm long; stigma minute, capitate. Fruit maturing into 1(–2) samaras, each samara 5.5–7 cm long, 1.7–2.5(–2.7) cm wide, the dorsal edge of the wing straight or slightly falcate, the ventral edge ovoid, the apex often apiculate, densely stellate-pubescent, but glabrescent in age; the seed-containing portion of the samara ellipsoid, 1.2–1.7 cm long, 0.5–0.7 cm in diameter, tuberculate, densely stellate-pubescent; the inner walls of the fruit densely pubescent, covered with ± 1 mm long, light golden-brown stellate and fasciculate hairs; seeds one per samara, ovoid, ± 1.3 mm long, ± 0.7 mm in diameter, apex attenuate, base circular, flattened, seed coat dark brown, finely reticulate, glabrous. Fig. 10, p. 58.

TANZANIA. Handeni District: fl. fr. 15 Feb. 1948, *Mohamedi* AH 9877! & Kideleka Mission, near Handeni, fl. 16 Feb. 1948, *Bally* 5808! & Kwaluwala, 13 km from Handeni, fr., 17 Aug. 1961, *Semsei* 3244!
DISTR. **T** 3; endemic, known only from the type locality and vicinity
HAB. In forest, wooded grassland, or deciduous thickets of *Acacia, Commiphora*, and succulent *Euphorbia*, also in native cultivations (fide *Mohamedi* AH 9877); ± 650 m

LOCAL USES. Bathing in an infusion of the bark is reputed to be a cure for scabies (*Mohamedi* AH 9877)
CONSERVATION NOTES. In view of the fact that this species is only known from the vicinity of the type locality it is here assessed as CR D1 (critically endangered).
NOTE. While this account was in proof, Quentin Luke brought to light two sterile specimens of what appear to be this taxon, or another, closely related. The first, *Alweid and Frontier Team* in YSA 268S (Mt Kanga, **T** 6, 26 April 2006) is from a 10 m tree and is a very good match for a *Mansonia* in leaf shape, margin, variation and indumentum. The second *Luke et al.* 8686 (Udzungwa Mts, **T** 7, 3 June 2002) is from a 1 m sapling, initially determined as *Buchnerodendron* (Flacourtiaceae). Its leaves differ from other *Mansonia* in being palmately lobed and much larger (to 32 × 40 cm) and in having only 1–2-armed hairs. However a similar pattern is seen in other tree taxa in this account: adults having simple, unlobed leaves, while juveniles have larger, palmately lobed leaves. *Cola scheffleri* is an example of a tree species formerly restricted to **T** 3 until it was recently discovered at Kanga Mt and the Udzungwas. Fertile material is needed of *Mansonia* from **T** 6 or **T** 7 to be sure of the identification at species level (Martin Cheek).

8. DOMBEYA

Cav., Diss. 2, App.:[4](1786) nom. cons.; Seyani, *Dombeya* (1982) & *Dombeya* in Africa (1991)

Walkuffa Bruce, Travels to discover the source of the Nile, 5: 67, fig. (1790)
Xeropetalum Delile Cent. Pl. Afr. Voy. Merôe: 84 (1826)

Trees or shrubs, deciduous when flowering, more rarely evergreen, pubescent with stellate, simple and glandular hairs. Leaves petiolate; blade 3–5(–7)-palmately lobed, or entire, base often cordate, margin serrate, dentate, or rarely crenate; stipules narrowly triangular, caducous. Inflorescence axillary, pedunculate, cymose, bracteate, subumbellate or dichasial with 10–40 flowers, pedicellate; or epicalycular bracts whorled, usually equal, caducous, rarely alternate on pedicel. Sepals 5, valvate, united at base, triangular, often reflexed, nectariferous adaxially at base, outer surface pubescent. Petals 5, contorted, spreading, white, sometimes red-veined or based, or pink, asymmetrical, dimidiately ovate or elliptic, shortly clawed, glabrous, brown-marcescent. Androgynophore absent; stamens in a single whorl, filaments united in a short tube (but 1.5–1.6 cm long in *D. amaniensis*) at base, fertile stamens 10–15, in groups of 3 alternating with 5 slightly longer or subequal ligulate-spatulate petaloid staminodes; anthers dehiscing by slits; ovary ovoid to oblate, tomentose or rarely glandular, with 3–5 locules each with 2–8 collateral ascending ovules, placentation axile, style ± as long as stamens, shortly (2–)3–5-branched. Fruit spherical to ovoid, brown, dry, slightly woody, dehiscing loculicidally by 3–5 valves. Seeds 1–several per locule, subreniform or angular, brown or black, not fleshy.

A genus of about 200 species confined to continental Africa (some 18 species), Madagascar (some 180 species), Mascarenes, and the Arabian Peninsula.

Two hybrids and several species, including the E African *D. burgessiae* are sometimes cultivated throughout the tropics and subtropics as ornamental garden plants, although not in E Africa. Most E African species are used for the fibre of their bark, and their wood is used for bows, tool handles or firewood.

This treatment is based on Seyani's 1982 validly published doctoral thesis revision ("A taxonomic study of *Dombeya* Cav. (*Sterculiaceae*) in Africa with special reference to species delimitation"). This was published in 1991 in modified form ("Dombeya in Africa") as the second book in the series Opera Botanica Belgica. In Seyani's works, many species formerly accepted are reduced to synonymy. Further work might show that some of these might be worth maintaining at one infraspecific rank or another.

1. Petals mostly > 1.2 cm long; style 5-branched 2
   Petals mostly < 1.2 cm long (to 1.7 cm in *D. quinqueseta*); style 2–3-branched 6
2. Inflorescence subumbellate, 4–5 cm long; peduncle 0.4–1 cm long, petals 3.1–4.7 cm long .   2. *D. amaniensis* (p. 61)
   Inflorescence corymbose, dichomotously cincinnate (scorpioid cyme) or with two subumbellate branches 4–27 cm long; peduncle 2–21 cm long; petals 0.9–2.9 cm long 3
3. Inflorescence dichomotously cincinnate (scorpioid cyme); flowers distichous on the two branches; the largest epicalyx lobes ± as long as wide, apex usually rounded . . . .   1. *D. acutangula* (p. 61)
   Inflorescence corymbose or with two subumbellate branches; epicalyx lobes much longer than wide, apex usually acuminate or acute (but rounded in some *D. buettneri*) 4
4. Leaves unlobed, very rarely obscurely lobed, margin with 4–10 teeth per 1 cm . . . .   5. *D. torrida* (p. 66)
   Leaves obscurely to distinctly 3–5-lobed, margin with 1–4 teeth per 1 cm 5
5. Leaf base shallowly to deeply cordate, the two lobes 90° or more apart; apex acute to rounded or emarginate . . . .   3. *D. buettneri* (p. 63)
   Leaf base deeply cordate, the two lobes overlapping or nearly so; apex acuminate . . . .   4. *D. burgessiae* (p. 64)
6. Inflorescences in fascicles subtended by scale leaves, borne on condensed, leafless, specialized short shoots usually arising from the axils of fallen leaves 7
   Inflorescences single, in axils of leaves at the uppermost 4–6 nodes of normal leafy shoots 9

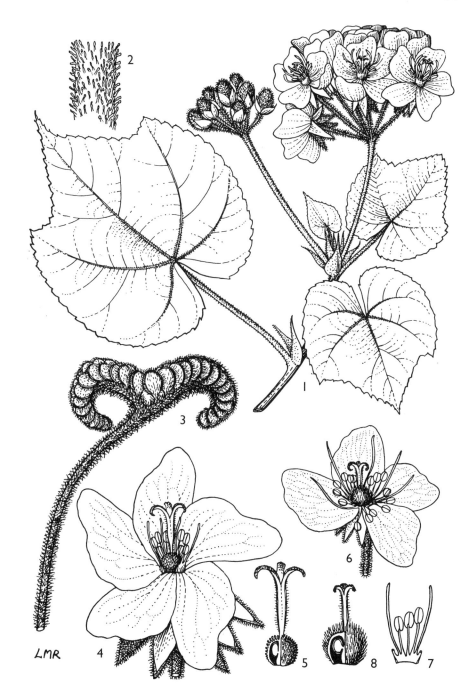

FIG. 11. *DOMBEYA BURGESSIEAE* — **1**, habit × ²/₃; **2**, stem indumentum; *DOMBEYA ACUTANGULA* — **3**, young inflorescence × 2; *DOMBEYA SHUPANGAE* — **4**, flower with part of staminal tube removed × 4; **5**, gynoecium with part of ovary wall removed × 6; *DOMBEYA ROTUNDIFOLIA* — **6**, flower with one petal removed × 3; **7**, part of androecium × 3; **8**, gynoecium, with part of ovary wall removed × 3. 1–2 from *Richards* 5573; **3** from *Riddelsdell* 95; 4–5 from *Brass* 17577; 6–8 from *Plowes* 1631. Reproduced from F.Z. 1: t. 98. Drawn by Lura Mason Ripley.

7. Ovary indumentum shortly golden glandular (but
  some non-glandular hairs sometimes at style base)   6. *D. shupangae* (p. 67)
  Ovary indumentum white, long, non-glandular 8
8. Stems pale brown; leaf apex acuminate or
  subacuminate; margin crenate-serrate; petals
  10–16.5 mm long . . . .   7. *D. quinqueseta* (p. 68)
  Stems purple-brown; leaf apex rounded or rarely
  acute; margin dentate; petals 8–14 mm long . . . . .   8. *D. rotundifolia* (p. 69)
9. Leaves mostly obovate, rarely suborbicular, apex
  rounded, rarely acute or obtuse; petals 5–9 mm
  long; **K** 7 & **T** 3 . . . .   10. *D. taylorii* (p. 72)
  Leaves ovate to suborbicular, apex ± acuminate,
  rarely rounded; petals 7–13 mm long . . . .   9. *D. kirkii* (p. 71)

1. **Dombeya acutangula** *Cav.*, Diss. Bot. 3: 123, t. 38, f. 2 (1787); Seyani, Dombeya: 84 (1982) & Dombeya in Africa: 59 (1991). Type: Reunion, *Commerson* (P-JU, holo.)

Small tree or shrub 2–3(–8)m tall; stems pubescent, tomentose or glabrescent, hairs simple, stellate, and minute, subsessile-glandular. Leaves ovate to broadly ovate, 6–20 cm long, 5–13(–17) cm wide, dimorphic; juvenile leaves deeply 5–9-lobed; adult leaves shallowly 3–5-lobed/angled or unlobed, apex acuminate, margin dentate or crenate-serrate, base truncate to shallowly cordate, upper surface scabridulous, lower surface softly pubescent to tomentose rarely subglabrous; petiole 2–13 cm long; stipules obliquely lanceolate, 0.5–1.5 cm long, 0.15–0.7 cm wide. Inflorescence axillary, 5–16 cm long, dichotomously cincinnate, many-flowered; peduncle 3.7–12.5 cm long, tomentose or densely pubescent, peduncle bracts absent; branches 2.2–9.4 cm long, unbranched, slender; pedicels distichous, 0.7–1.9 cm long; epicalyx bracts at base of calyx tube, rarely on it, dimorphic: the largest broadly ovate or suborbicular, apex rounded or obtuse, base cordate to auriculate, sessile, the two smaller bracts 0.5–0.9 cm long, 0.3–0.5 cm wide, base stipulate. Calyx lobes reflexed, lanceolate, 0.7–0.8 cm by 0.2–0.3 cm, acuminate, appressed-tomentose outside. Petals spreading wide, white, 1–1.9 cm long, 0.5–1.8 cm wide; androecium 0.6–1 cm long. Staminal tube widening towards apex, 0.1–0.2 cm long; staminodes 0.6–0.8 cm long; longest stamen 0.3–0.8 cm; anthers 0.15–0.3 cm long, apiculate. Ovary globose, tomentose, style ± 0.75 cm long, 5-branched, glabrous. Capsule subwoody, oblate, 0.4–0.6 cm diameter, 5-ribbed or 5-winged, tomentose; seeds 1.2 × 0.8 mm, dark brown. Fig. 11: 3, p. 61.

Tanzania. Handeni District: Kwa Mkono, June 1966, *Archbold* 738!; Pangani District: Karoti, Mkwaja, June 1956, *Tanner* 2936!; Mpwapa District: Mlali, May 1961, *Akiley* 79; Zanzibar, Haikajwa, June 1930, *Vaughan* 1368!
Distr. **T** 2–8; **Z**; Zambia, Malawi, Mozambique; Madagascar, Mascarene Islands
Hab. Forest, woodland, bushland or thicket; 0–1500 m

Syn. *D. cincinnata* K.Schum. in P.O.A. C: 170 (1895); T.T.C.L. 595 (1949); Wild in F.Z. 1: 520, t. 98 f. b (1961); Wild & Gonç. in Fl. Moçamb. 27: 4 (1979). Type: Tanzania, Tanga District: Amboni, *Holst* 2204 (B†, holo.; BM, K!, iso.)
    D. stuhlmannii K.Schum. in N.B.G.B. 2: 302 (1899). Type: Tanzania, Zanzibar, Ukwira, Kissema, *Stuhlmann* 8414 (B†, holo.; BR, iso.)
    D. cincinnata K.Schum. var. *stuhlmannii* (K.Schum.) K.Schum. in E.M. 5: 32 (1900); T.T.C.L. 595 (1949)
    D. leucorrhoea K.Schum. in E.J. 30: 353 (1901). Type: Tanzania, Njombe District: Kinga, Rumbira, *Goetze* 898 (B†, holo.; BM, BR, K! iso)

Conservation notes. A widespread and common species, here assessed as of Least Concern for conservation. The assessment of this species as critically endangered (Tezoo, V. & Strahm, W. 2000. *Dombeya acutangula*. In: IUCN 2004. 2004 IUCN Red List of Threatened Species) is erroneous since based on the assumption that the taxon is confined to Mauritius.

2. **Dombeya amaniensis** *Engl.* in E.J. 39: 581 (1907) & in V.E. 3 (2): 428, t. 198 (1921), T.T.C.L: 595 (1949); Seyani, Dombeya: 93 (1982) & Dombeya in Africa: 64 (1991).Type: Tanzania, Lushoto District: Amani, road to Sigi, *Braun* s.n. in Herb. Amani 668 (B†, syn.; EA, lecto., chosen by Seyani (1991) and locality unclear, *Engler* 3413a (B†, syn.)

Small tree or shrub 3–5 m tall; stems angular at nodes, densely pubescent to glabrescent, hairs simple, stellate and subsessile-glandular. Leaves suborbicular, 20–23 cm long, 23–26 cm wide, shallowly 3–5-lobed, apex acute, margin dentate-crenate, base deeply cordate, upper and lower surface densely pubescent; petiole 18–27 cm long; stipules caducous, lanceolate, 1.2–1.6 cm long, 0.5–0.6 cm wide, acuminate, tomentose. Inflorescence axillary, 4–5 cm long, subumbellate, 4–7-flowered; peduncle 0.4–1 cm long, tomentose; bracts not seen; branches absent; pedicels 0.6–1.2 cm long; epicalyx bracts at calyx base, narrowly lanceolate, 2–2.4 cm long, 0.3 cm wide, long-acuminate, sparsely puberulous, persistent. Calyx lobes erect, narrowly triangular, 3–3.5 cm long, 0.4–0.5 cm wide, acute, stellate-pilose. Petals 3.1–4.7 cm long, (1.9–)2.6–3.2 cm wide; androecium 2.6–3 cm long; staminal tube cylindrical (0.9–)1.5–1.6 cm long, staminodes ± 1.6 cm long, longest stamen ± 1.9 cm long; anthers 0.4–0.6 cm long. Ovary subglobose, tomentose, style 2.1–2.9 cm long, 5–6-branched, pubescent. Fruit and seeds unknown.

TANZANIA. Masai District: Makuyuni, *Koritschoner* 816; Lushuto District: Burwa Tea Estate, Aug. 1961, *Ali Omari* in *Richards* 153444!; Tabora District: Mahenge, May 1932, *Schlieben* 2169!
DISTR. **T** 2, 3, 6, 7; not known elsewhere
HAB. Marshy places in forest, by rivers, and in wooded grassland; 300–1000 m

CONSERVATION NOTES. Known from a few collections at five localities (area of occupancy estimated to be less than 500 km²); its habitat is threatened with a decline in quality at one of these localities by ongoing wood extraction (Luke and Vollesen, pers. comm. 2005). The taxon is accordingly here assessed as EN B2a, b(iii), i.e. Endangered. Further, on the ground assessment is needed at all four sites and may result in the species being assessed as Critically Endangered. Previously the taxon was assessed as VU B1+2b on the basis of a smaller altitudinal and geographical range (Lovett, J. & Clarke, G.P. 1998. *Dombeya amaniensis*. In: IUCN 2004. Red List of Threatened Species).
NOTES. *D. amaniensis* is unlikely to be confused with any other species. The shortly peduncled subumbellate inflorescence with only 4–7 very large flowers is unique in African *Dombeya*, as is the long (1.5–1.6 cm) staminal tube. The record from **T** 7 is based on Luke (pers. comm) regarding *Luke & Luke* 4869 & 8745 (EA n.v.) from Udzungwa MNP. Luke also gives a sight record (outside Matundu FR) to support the lower altitudinal range.

3. **Dombeya buettneri** *K.Schum.* in E.J. 15: 133 (1892) & in E.M. 5: 27, t. 2 (1900); F.W.T.A. ed. 2, 1(2): 317 (1958); Seyani, Dombeya: 96 (1982) & Dombeya in Africa: 75 (1991); Vollesen in Fl. Eth. 2, 2: 168 (1995). Type: Togo, Kentschenki stream, *Buettner* 347 (B†, syn., K!, lecto., chosen by Seyani, 1991)

Small tree or shrub 1.5–6(–9) m tall; stems terete or angled at nodes, densely pilose, tomentose, tomentellous or glabrescent, hairs simple, stellate and subsessile (rarely stalked) glandular. Leaves suborbicular, obscurely to distinctly 3–5-lobed, 6–21(–24) cm long, (4–)6–24 cm wide, apex acute to rounded or emarginate, margin coarsely crenate-serrate, base shallowly to deeply cordate, upper surface softly pubescent or puberulous, lower surface densely pilose-pubescent or tomentose; petiole 2–18 cm long; stipules narrowly lanceolate to broadly ovate, 0.6–2 cm long, 0.2–1.1 cm wide, acuminate, persistent. Inflorescence axillary, 4–23(–27) cm long, corymbose, rarely subumbellate, dense or lax, many-flowered; peduncle 2–21 cm long, short or slender, pilose to tomentellous; bracts ovate-lanceolate, 0.6–1.2 cm long, 0.2–0.6 cm wide, acuminate, caducous; branches several, bracts as the peduncle-bracts, 0.5–0.9 cm long; pedicels 0.7–2.1 cm long; epicalyx bracts at base of calyx or inserted on pedicel, ovate, 0.6–1.6 cm long, 0.2–0.7 cm wide, acute to acuminate, caducous. Calyx lobes reflexed or erect, lanceolate-acuminate, 0.6–1.4 cm long, 0.2–0.5 cm wide, pubescent outside. Petals white, sometimes with a red base, 0.9–1.9 cm long, 0.6–1.6 cm wide.

Androecium 0.9–1.6 cm long; staminal tube 0.3–0.7 cm long, outer surface strongly convex; staminodes 0.5–1.1 cm long; largest stamen 0.4–0.6 cm long; anthers 0.2–0.4 cm long. Ovary globose, tomentose; style 0.4–0.9 cm long, filiform or stout, pubescent or glabrous, 5-branched. Capsule subglobose or ovoid, 0.7–0.9 cm diameter, appressed silky hairy; seeds 0.2–0.3 cm long, 0.1–0.3 cm wide, dark brown.

UGANDA. West Nile District: Niagak R., Dec. 1931, *Brasnett* 303!; Ankole District: Igara, May 1939, *Purseglove* P 679!; Mubende District: North Mubende, Dec. 1937, *Sitwell* in *Eggeling* 3476!
TANZANIA. Bukoba District: Kabirizi, Oct. 1931, *Haarer* 2231!; Mpanda District: Mwese, May 1975, *Kahurananga* et al. 2597!; Kigoma District: Mahari Mts, Ujamba, July 1958, *Mgaza* in *Jefford et al.* 177!
DISTR. U 1, 2, 4; T 1, 4; Guinea, Sierra Leone, Ivory Coast, Ghana, Togo, Nigeria, Cameroon, Central African Republic, Congo-Kinshasa, Rwanda, Burundi, Sudan, Zambia
HAB. Grassland, bushland, wooded grassland and open woodland or forest; 1100–1800(–2200) m

SYN. *D. pedunculata* K.Schum. in P.O.A. C: 269 (1895); E.M. 5: 27, t. 2 (1900); T.T.C.L. 595 (1949). Type: Tanzania, Bukoba District: 'Kafuro in K', *Stuhlmann* 1754 (B†, holo.); neotype: Burundi, Lacs Edouard et Kivu, *Fries* 1510 (UPS, neo., chosen by Seyani, 1991, but see Note below)
     D. bagshawei Bak.f. in J.L.S. 37: 127 (1905); T.T.C.L.: 594 (1949); I.T.U.: 416 (1952); Bamps in F.C.B. 10: 244 (1963). Type: Uganda, Ankole District: Irunga Hill, *Bagshawe* 391 (BM, holo.)
     D. seretii De Wild. in Ann. Mus. Congo Belge, Bot., Ser. 5, 2: 50 (1907); Bamps in F.C.B. 10: 243. Type: Congo-Kinshasa, Suronga, *Seret* 358 (BR, holo.)
     D. mildbraedii Engl. in E.J. 45: 317 (1910). Type: Uganda, Ankole District: Mpororo, *Mildbraed* 347 (B†, holo.)
     D. quarrei De Wild., Contr. Fl. Kat. Suppl. 1: 49 (1927). Type: Congo-Kinshasa, Lumbumbashi [Elisabethville], *Quarré* 383 (BR, holo.; A, iso.)
     D. robynsii De Wild., Contr. Fl. Kat. Suppl. 1: 51 (1927); Bamps in F.C.B. 10: 245 (1963). Type: Congo-Kinshasa, Kisabi-Kasheka, *Robyns* 2075 (BR, lecto., chosen by Bamps, 1963, K, isolecto.)
     D. claessensii De Wild., Pl. Bequaert. 4: 415 (1928). Types: Congo-Kinshasa, Blukwa, *Claessens* 1438 & 1492 (BR, syn.)
     D. elskensii De Wild., Pl. Bequaert. 4: 414 (1928). Type: Burundi, Kitete, *Elskens* 28 (BR, holo.)
     D. emarginata E.A.Bruce in K.B. 1932: 94 (1932). Type: Uganda, Mubende, *Hargreaves* 2046 (K, holo.)
     D. kandoensis De Wild. & Staner, Contrib. Fl. Kat. Suppl. 4: 59 (1932). Type: Congo-Kinshasa, Kando, *De Witte* 208 (BR, holo.)

CONSERVATION NOTES. A widespread and common species, here assessed as of Least Concern.
NOTES. The protologue of *D. pedunculata* K.Schum. mentions only the single specimen cited above; Seyani (1982) erroneously cites this and three other specimens as syntypes. One of these, *Fries* 1510, he designates as lectotype. Later Seyani (1991) redesignated *Fries* 1510 as a neotype. Since the selection of this specimen was based on an error, it is possible that a better choice is available.

4. **Dombeya burgessiae** *Harv.*, Fl. Capensis 2: 590 (1862); Mast., F.T.A. 1: 228 (1868); K. Schum. in E.M. 5: 28 (1900); Wild in F.Z. 1: 522 (1961); Bamps in F.C.B. 10: 246 (1963); Seyani, Dombeya: 104 (1982) & Dombeya in Africa: 66 (1991); Blundell, Wild Fl. E Africa: t. 52 (1987); K.T.S.L. 163, map (1994). Types: South Africa, Zululand, *McKen* s.n. (TCD, syn.); Klip River, Natal, *Gerrard* s.n. (TCD, lecto., chosen by Seyani, 1988; K, isolecto.)

Small tree or shrub 2–4(–8) tall; stems terete, densely pubescent to glabrescent, hairs simple, stellate, and subsessile- and stalked-glandular. Leaves suborbicular, distinctly 3-lobed, rarely 5-lobed or entire, 11–23 cm long, 7–18 cm wide, apex acuminate, margin coarsely and irregularly crenate or dentate, base deeply cordate, upper surface densely pubescent with long-armed stellate hairs; petiole 7–15 cm long; stipules lanceolate to broadly ovate, 0.8–1.3 cm long, 0.3–0.6 cm wide, long-acuminate. Inflorescence axillary, 6–21 cm long, corymbose or ± umbellate, many-flowered, peduncle 3.8–16.5 cm long, sparsely to densely pubescent; bracts ovate, 1–1.2 cm long, acuminate, pubescent or glabrescent, caducous; branches absent or condensed, branch-bracts not seen; pedicels 2.2–3.8 cm long; epicalyx bracts at calyx base, rarely on the pedicel, elliptic, ovate, oblong, lanceolate or oblanceolate, 1–1.4 cm long, 0.2–0.5 cm wide, acute to long-acuminate, caducous. Calyx lobes reflexed, lanceolate, 0.9–1.3 cm long, 0.2–0.8 cm wide, acute, pubescent outside. Petals spreading, pink or white, with or without a red base or veins, 1.2–2.5 cm long, 1–2.5 cm wide. Androecium 1.2 cm long; staminal tube 0.1–0.5 cm long, dilating towards apex or sides convex; staminodes 0.7–1.6 cm long; longest stamen 0.6–1 cm; anthers 0.2–0.5 cm long. Ovary globose, ovoid, tomentose, style 0.5–1.6 cm long, pubescent or glabrous, 5-branched. Capsule ovoid or globose, 0.5–1.5 cm in diameter, tomentose; seeds 0.3–0.4 × 0.2 cm, dark brown to black. Fig. 11: 1–2, p. 61.

UGANDA. Karamoja District: Mt Debasien, Jan. 1936, *Eggeling* 2835!; Mwanza District: Budongo, Busingiro Hill, Nov. 1933, *Eggeling* 1397; Masaka District: Kibula, Sep. 1945, *Purseglove* P1838!
KENYA. Turkana District: Murua Ngithigerr Mt, Dec. 1988, *Beentje, Powys & Dioli* 3915!; Ravine District: Sabatia, June 1941, *Gardner* 1221!; Masai District: Mara R., Ngerendei, March 1961, *Glover, Gwynne & Samuel* 227!
TANZANIA. Bukoba District: Bunazi, Sep. 1935, *Gillman* 581!; Mbulu District: Haraa Forest Reserve, May 1988, *Ruffo & Sigara* 3062!; Mbeya District: Mbeya, N Usafua Forest Reserve camp, Oct. 1947, *Brenan* 8192!
DISTR. **U** 1–4; **K** 2–6; **T** 1, 2, 4, 5, 7, 8; Congo-Kinshasa, Rwanda, Angola, Zambia, Malawi, Mozambique, Zimbabwe, Swaziland, South Africa
HAB. Grassland, wooded grassland, riverine thicket and scrub, open forest and forest edges; (900–)1500–2000(–2400) m

SYN. *D. mastersii* Hook.f., Bot. Mag. t. 5639 (1867); Mast., F.T.A. 1: 228 (1868); K. Schum. in E.M. 5: 26 (1900); T.T.C.L. 596 (1949). Type: cultivated, origin unknown
    D. tanganyikensis Baker in K.B. 1897: 244 (1897); K. Schum in E.M. 5: 40 (1900); Wild in F.Z. 1: 523, t. 98 f. A. (1961); Bamps in F.C.B. 10: 246 (1963). Type: Malawi, Fort Hill, 3,000-4,000 ft., *Whyte* s.n. (K, holo.)
    D. johnstonii Baker in K.B. 1898: 301 (1898); K. Schum. in E.M. 5: 33 (1900); T.T.C.L. 595 (1949); F.Z. 1: 523 (1961). Types: Malawi, Mpata–Tanganyika Plateau, *Whyte* s.n. (K, syn.); Nyika Plateau, *Whyte* s.n. (K, syn.)
    D. calantha K.Schum. in E.M. 5: 28 (1900); Sprague in Bot. Mag. t. 8424 (1912); Wild in F.Z. 1: 524 (1961). Type: Malawi, Zomba, *Whyte & McClounie* s.n. (K, holo.)
    D. lasiostylis K.Schum. in E.M. 5: 24 (1900); Wild in F.Z. 1: 524 (1961). Type: Malawi, Muata Manja, *Kirk* s.n. (B holo†; K iso.)
    D. parvifolia K. Schum. in E.M. 5: 30 (1900). Type: Uganda, Masaka District: Buddu, *Scott Elliot* 7475 bis (K, holo.)
    D. platypoda K.Schum. in E.M. 5: 29 (1900); T.T.C.L. 596 (1949). Type: Zambia, Tanganyika Plateau, Fwambo, *Carson* s.n. (K, holo.)
    D. kindtiana De Wild. in B.S.B.B. 40: 14 (1901). Type: Congo-Kinshasa, Lukafu, *Verdick* 542 (BR, holo.)
    D. auriculata K.Schum. in E.J. 30: 352 (1901); V.E. 3 (2): 427, t. 197 (1921); T.T.C.L. 595 (1949). Type: Tanzania, Usafua, Poroto Mts,*Goetze* 1285 (B†, holo.; BR, iso.)
    D. dawei Sprague in J.L.S. 37: 501 (1906); T.T.C.L. 595 (1949); I.T.U.: 418 (1951); F.P.U. 58 (1971). Type: Uganda, Buddu District, Masaka, *Dawe* 10 (K, holo.)
    D. nairobensis Engl. in E.J. 39: 583 (1907); I.T.U.: 419 (1951). Type: Kenya, Nairobi, *Kaessner* 963 (B†, holo.; BM, K, iso.)
    D. endlichii Engl. & K.Krause in E.J. 48: 551 (1912); T.T.C.L. 595 (1949). Type: Tanzania, Kilimanjaro, Kibo, *Endlich* 299 (B†, holo; W, iso.)
    D. velutina De Wild. & Staner, Contrib. Fl. Kat., Suppl. 4: 62 (1932). Type: Congo-Kinshasa, Marungu, *De Witte* 446 (BR, holo.)
    D. burttii Exell in J.B. 77: 165 (1939); T.T.C.L. 596 (1949). Type: Tanzania, Poroto [Mporoto] Mts, *Burtt* 6154 (BM, holo.; K, iso.)
    D. gamwelliae Exell in J.B. 70: 104 (1932). Type: Zambia, Abercorn, *Gamwell* 50 (BM, holo.)
    D. nyasica Exell in J.B. 77: 166 (1939); Wild in F.Z. 1: 523 (1961). Type: Malawi, between Dowa and Kota Kota, *Burtt* 6085 (BM, holo.; BR, K, iso.)
    D. trichoclada Mildbr. in N.B.G.B. 12: 196 (1934); T.T.C.L. 596 (1949). Type: Tanzania, Iringa, Lupembe, Ruhudje, *Schlieben* 1051a (B†, holo.; BR, iso.)
    D. globiflora Staner in Ann. Soc. Sci. Brux., ser. B, 50: 59 (1935). Type: Congo-Kinshasa, Haut-Katanga, *Quarré* 3294 (BR, holo.)
    D. greenwayi Wild in Bol. Soc. Brot., sér. 2, 3: 35 (1959); Wild in F.Z. 1: 521 (1961). Type: Zambia, Shiwa Ngandu, *Greenway* 5527 (EA; K holo.)
    D. sphaerantha Gilli in Ann. Naturhist. Mus. Wien, 74: 449 (1971). Type: Tanzania, Mbeya Mt, *Gilli* 351 (W, holo.)

CONSERVATION NOTES. A widespread and common species, here assessed as of Least Concern.

5. **Dombeya torrida** (*J.F.Gmel.*) *P.Bamps* in B.J.B.B. 32: 170 (1962); Hepper & Friis in Bot. Notis. 132: 397 (1979); Seyani, Dombeya: 130 (1982) & Dombeya in Africa: 80 (1991); K.T.S.L.: 165, map, fig. (1994); Vollesen in Fl. Eth. 2, 2: 168, fig. 80.2: 1–3 (1995). Type: Ethiopia, t. 20 in Bruce, 5 (1790)

Deciduous tree (7–)12–20 m tall; stems terete, hispid to glabrous, hairs simple, stellate, tufted and subsessile- (rarely stalked-) glandular. Leaves suborbicular, usually unlobed, 7–28 cm long, 5–20 cm wide, apex acuminate, margin finely and closely serrate, base deeply cordate, upper surface scabridulous, rarely glabrous, lower surface puberulous or tomentellous, frequently with tufts of hairs at base of main nerves; petiole 3.6–20 cm long; stipules with the basal half transversely oblong or ellipsoid-oblong, the upper half narrowly triangular, 0.5–2.1 cm long, 0.1–0.8 cm wide, caducous to persistent. Inflorescence axillary, 5.5–22 cm long, with two or more branches, each sub-umbellate, many-flowered; peduncle 2.3–13.4 cm; bracts lanceolate, 0.7–1.7 cm long, 0.3–0.6 cm wide, acuminate, glabrescent, caducous; branches usually once-forked near the apex, bracteoles 1–1.2 cm long; epicalyx bracts inserted at calyx base, ovate, 0.4–1.1 cm long, 0.2–0.5 cm wide; apex acute to acuminate, caducous. Calyx lobes erect or reflexed, lanceolate-triangular, 0.5–1.3 × 0.2–0.4 cm. Petals pink or white, with or without a red base or veins, 1.1–1.9 cm long. Androecium 0.8–1.8 cm long; staminal tube dilating towards apex or sides convex; staminodes 0.4–1.3 cm long; longest stamen 0.6–1.1 cm; anthers 0.2–0.3 cm long. Ovary globose or ovoid, tomentose; style 0.6–1.1 cm long, 5-branched. Capsule globose, rarely depressed–ovoid, tomentose; seeds ovoid-oblong, 0.3 × 0.2 cm, dark brown.

subsp. **torrida**

Leaves abruptly narrowed to the acumen, lower surface densely puberulous with a mixture of long and short armed stellate hairs.

UGANDA. Karamoja District: Mt Moroto, Feb. 1936, *Eggeling* 2883!; Kigezi District: near Behungi, Oct. 1929, *Snowden* 1517!; Mbale District: Elgon, Bulambuli, Nov. 1933, *Tothill* 2358
KENYA. Northern Frontier District: Mt Nyiro, July 1960, *Kerfoot* 1941!; Nakuru District: Eastern Mau Reserve, June 1957, *Trapnell* 2310!; Nyandarua/Aberdares, Kinangop side, Sep. 1963, *Verdcourt* 3782!
TANZANIA. Masai/Mbulu District: Ngorongoro Conservation area, Empakaai Crater, June 1973, *Frame* 185!; Arusha District: Mt Meru, *Bancroft* 50!; Moshi District: Moshi District: Marangu, Oct. 1960, *Steele* 64!
DISTR. **U** 1–3; **K** 1–3, 4, 5?, 6; **T** 2; Congo-Kinshasa, Rwanda, Burundi, Sudan, Eritrea, Ethiopia, Djibouti; Yemen
HAB. Open forest and forest edges, often with *Juniperus, Hagenia, Podocarpus* or bamboo, woodland, or in pasture; (1700–)2100–3050 m

SYN. *Walcuffa torrida* J.F. Gmel., Syst. Nat: 1029 (1791)
　　 Dombeya bruceana A.Rich., Tent. Fl. Abyss. 1: 77 (1847); Mast., F.T.A. 1: 229 (1868); K. Schum., E.M. 5: 22 (1900). Types: Ethiopia, Tigray, Addischoa, *Schimper* 378 (P, syn., BM, BR, RD, UPS, K, isosyn.); Tigray, Shollon, *Quartin-Dillon* s.n. (P, syn.), *Petit* s.n. (P, syn.)
　　 D. schimperiana A.Rich., Tent. Fl. Abyss. 1: 78 (1847); Mast., F.T.A. 1: 229 (1868). Type: Ethiopia, Abera, *Schimper* 845 (P, holo., BR, K, FT, RD, iso.).
　　 D. goetzenii K.Schum. in Von Goetzen, Durch. Afr. von O. nach W.: 6 & 24 (1895) & in E.M. 5: 24 (1900); I.T.U.: 418 (1951); Bamps in F.C.B. 10: 242 (1963); Hamilton, Uganda For. Trees: 118 (1981). Type: Congo-Kinshasa, Kirunga, *Goetzen & Prittwitz* 97 (B†, holo.)
　　 D. schimperiana A.Rich. var. *glabrata* K.Schum. in Von Goetzen, Durch Afr. von O. nach W.: 23 (1895). Type: Ethiopia, Abera, *Steudner* 1157 (B, holo.†; K, P, iso.)
　　 D. runsoroensis K.Schum. in E.M. 5: 23 (1900). Types: without locality details, *Stuhlmann* 2736 (B†, syn.); Congo-Kinshasa, without locality details, *Stuhlmann* 2347 (B†, syn.)
　　 D. leucoderma K.Schum. in P.O.A. C: 270 (1895)& in E.M. 5: 21 (1900); T.T.C.L. 596 (1949). Type: Tanzania, Marangu, *Volkens* 1296 (B†, holo.; K, lecto., chosen by *Seyani* 1982)

D. albiflora K.Schum. in E.J. 33: 308 (1903). Type: Ethiopia, Schoa und Galla, Mandagarrha, *Ellenbeck* 1632 (B†, holo.)

D. faucicola K.Schum. in E.J. 34: 323 (1904); Engl. in V.E. 3(2): 426, t. 196 (1921). Type: Kenya, Nakuru, *Engler* 2048 (B†, holo.)

D. elliotii K.Schum. & Engl. in E.J. 39: 582 (1907). Type: Kenya, Masai Highlands, *Elliot* 19 (B†, holo.; K., iso.)

D. gallana K.Schum. & Engl. in E.J. 39: 583 (1907). Type: Ethiopia, Harar, Gara Mulata, *Ellenbeck* 558 (B†, holo.)

D. niangaraensis De Wild. Ann. Mus. Congo. Belge, Bot., sér. 5 (2): 49, t. 11 (1907). Type: Congo-Kinshasa, Niangara-Gambara, *Seret* 358 bis (BR, holo.)

D. bequaertii De Wild., Pl. Bequaert. 1: 510 (1922). Type: Congo-Kinshasa, Rutshuru, *Bequaert* 5918 (BR, holo.)

D. bogoriensis De Wild., Pl. Bequaert. 1: 511 (1922). Type: Congo-Kinshasa, Lake Albert, Bogoro, *Bequaert* 4967 (BR, holo.)

D. ruwenzoriensis De Wild., Pl. Bequaert. 1: 513 (1922). Type: Congo-Kinshasa, Ruwenzori, *Bequaert* 3642 (BR, holo.)

D. heterotricha Mildbr. in N.B.G.B. 12: 195 (1934); T.T.C.L. 596 (1949). Type: Tanzania, Kilimanjaro, Rongai, *Schlieben* 5052 (B†, holo.; BR, MAD, iso.)

D. stipulosa Chiov. in Atti R. Acad. Ital. Mem. Cl. Fis. 11: 21 (1940). Type: Ethiopia, Welega, Sajo,*Giordano* 2471 (FT, holo.)

CONSERVATION NOTES. A widespread and common taxon, here assessed as of Least Concern.
NOTES. Silviculturally important as the first tree species to recolonize where forest cover has been destroyed. Usually very common locally (*Battiscombe* 564).

subsp. **erythroleuca** (*K.Schum.*) *Seyani*, Dombeya in Afr. 2: 85 (1991). Type: Tanzania, Njombe District: Kinga Mts, Bulongwa, *Goetze* 927 (B†, holo.; BM, lecto., chosen by Seyani 1991; BR, iso.)

Leaves gradually narrowed to the acumen; lower surface sparsely puberulous, mainly with short-armed stellate hairs.

TANZANIA. Lushuto District: Shume-Magamba Forest, July 1954, *Hughes* 204!; Kilosa District: Ukaguru Mts, Ikwamba above Tengeta, Aug. 1972, *Mabberley & Salehe* 1486!; Songea District: Lupembe Hill, May 1956, *Milne-Redhead & Taylor* 10524!
DISTR. **T** 2–4, 6–8; Zambia and Malawi
HAB. As for *D. torrida* subsp. *torrida*

SYN. *D. erythroleuca* K.Schum. in E.J. 30: 353 (1901); T.T.C.L. 596 (1949); Wild in F.Z. 1: 525 (1961)
     D. macrotis K.Schum in E.J. 33: 309 (1903); T.T.C.L. 596 (1949). Type: Tanzania, Morogoro District: Uluguru,, Lukwangule, *Stuhlmann* 9130 (B†, holo.)
     D. malacoxylon K.Schum. in E.J. 33: 309 (1903); T.T.C.L. 597 (1949). Type: Tanzania, Lushoto District: Usambaras, Kwai, *Eick* 63 (B†, holo.)
     D. schoenodoter K.Schum. in E.J. 33: 310 (1903); T.T.C.L. 597 (1949).Type: Tanzania, Lushoto District: Usambaras, Kwai, *Albers* 13 (B†, holo.)
     D. monticola K.Schum. in E.J. 34: 324 (1904); T.T.C.L. 597 (1949). Type: Tanzania, Lushoto District: Usambaras, Kwai, *Engler* 1260 (B, holo.)
     D. sisyrocarpa Gilli in Ann. Naturhist. Mus. Wien, 74: 446 (1971). Type: Tanzania, District unclear, Tchenzema, *Gilli* 350 (W, holo.)

CONSERVATION NOTES. A widespread and common taxon, here assessed as of Least Concern.

6. **Dombeya mupangae** *K.Schum.* in E.M. 5:39 (1900); Wild in F.Z. 1: 526, t. 98 (1961); Seyani, Dombeya: 184 (1982) & Dombeya in Africa: 109 (1991). Type: Mozambique, Chupanga, *Kirk* s.n. (K!, holo.)

Tree 3–8 m; stems terete, purple-brown, pubescent to glabrescent, hairs simple, stellate, and subsessile-glandular. Leaves orbicular, sometimes ovate, entire or slightly 3-lobed, 10–26 cm long, 9–24 cm wide, apex rounded to subacuminate, margin entire or shallowly and inconspicuously coarsely crenate, base cordate, the two lobes overlapping or up to 120° apart, upper surface scabrid or subscabrid, lower surface felty; petiole 3.5–17 cm long; stipules lanceolate, 0.2–0.6 cm long, 0.1–0.2 cm wide, early caducous. Inflorescences 2–5 borne on condensed (usually 0.2–1 cm long) leafless specialised short shoots arising from the axils of usually fallen leaves near the apex of otherwise leafy stems, 5–15 cm long; scale-leaves brown, papery, ovate, 3–4 mm long; peduncle 1.5–7.5 cm long; bracts caducous, not seen; branches forking 2–3 times; bracteoles not seen; pedicels 0.6–2 cm long, villose to subglabrous; epicalyx bracts inserted at calyx base, elliptic or subspatulate, 0.1–0.2 cm long, early caducous. Calyx lobes reflexed, lanceolate, ± 0.6 cm long, 0.15 cm wide, pubescent. Petals spreading, opening white, often turning pink, 0.7–1.3 cm long, 0.5–0.6(–1) cm wide. Androecium 0.5–0.7 cm long; staminal tube cylindrical, 0.1 cm long; staminodes 0.5–0.6 cm long; longest stamen 0.4–0.5 cm, anthers ± 0.2 cm long. Ovary globose, 3-lobed, sparsely covered with short yellow simple glandular hairs ± 0.1 mm long, sometimes with a few short white stellate hairs at the style, rarely with white stellate hairs extending over the ovary; style 0.2–0.4 cm long, glabrous, 2–3-branched. Capsule globose to oblate, 0.6–0.7 cm diameter, indumentum as ovary; seeds ± 0.3 cm long, 0.2 cm wide, dark brown. Fig. 11: 4–5, p. 61.

subsp. **mupangae**

Inflorescence axes villose, stellate hairs with white, erect arms, ± 1 mm long.

KENYA. Teita District: Bura, date unknown, *Sacleux* 1995
TANZANIA Lushuto District: Kwembago village, 14 Aug. 1970, *Mshana* 113; Kilosa District: Mandege-Uponela, 1 Aug. 1972, *Mabberley* 1321!; Lindi District: Nachingwea, 14 Aug. 1953, *Anderson* 927; Zanzibar, 1839, *Burton* s.n.!
DISTR. **K** 7; **T** 3–8, **Z**; Congo-Kinshasa, Malawi and Mozambique
HAB. Secondary bushland and wooded grassland, less usually in forest and at forest edges; 0–1500 m

SYN. *D. katangensis* De Wild. & Th.Dur. in B.S.B.B. 40:13 (1901). Type: Congo-Kinshasa, Lukafu, *Verdick* 20 (BR, holo.)
    D. shupangae K.Schum. var. *katangensis* (De Wild. & Th.Dur.) Bamps in F.C.B. 10: 252 (1963)

CONSERVATION NOTES. A widespread and common taxon, here assessed as of Least Concern.
NOTES. *Dombeya mupangae* is unique among E African *Dombeya* in its ovary indumentum. It has short, yellow, simple, glandular hairs rather than the long white stellate indumentum that conceals the ovary in all other species. It is most closely related to *D. rotundifolia* and *D. quinqueseta*. These three species all produce inflorescences from very short specialised, lateral, leafless shoots borne in the axils of the fallen leaves of stems of the previous season's growth. In *D. mupangae*, however, the specialised shoots are usually produced only from the top 4–5 nodes, compressed into the apical 2–3 cm of these older stems. This results in such stems appearing to bear a single flat-topped terminal inflorescence ± 30 cm wide. Moreover, the stem below these pseudo-inflorescences retain old leaves from the previous season. In contrast, the other species mentioned tend to bear the specialized shoots along the length of the stems, which are usually completely leafless.
    Subspecies *glabrescens* (Bamps) Seyani (comb. ined.!) is distinguished by its glabrescent inflorescence axes and is restricted to Shaba Province, Congo-Kinshasa (Seyani op.cit. 1982: 189). Although not maintained later by Seyani (op. cit. 1991) it seems to me to merit recognition. It is arguable that Bamps earlier designation of varietal status is more appropriate.
    Schumann originally published the species epithet as *mupangae*. Sprague (J. Bot. 59: 349 (1921)) corrected this to *shupangae* on the basis that Schumann's intention was to commemorate the locality of the type collection that he, Sprague, gives as Shupanga. Subsequent authors have followed him. The name on the type specimen at Kew is written as Chupanga!

7. **Dombeya quinqueseta** (*Delile*) *Exell* in J.B. 73: 263 (1935); F.W.T.A. ed. 2, 1(2): 317 (1958); Bamps in F.C.B. 10: 249 (1963); Seyani, Dombeya: 163 (1982) & Dombeya in Africa: 105 (1991); K.T.S.L.: 164 (1994); Vollesen in Fl. Eth. 2, 2: 170 (1995). Type: Sudan, *Cailliaud* s.n. (MPU, holo.)

Small tree or shrub to 6 m; stems terete, tomentose to glabrous, hairs simple, stellate and subsessile-glandular; older stems becoming corky, pale brown. Leaves suborbicular, rarely shallowly 3-lobed, 12–24 × 10–21 cm, shortly acuminate, margin distinctly crenate-serrate, base deeply cordate, rarely crenulate-denticulate, upper surface scabridulous, lower surface tomentose to minutely puberulous; petiole 5–17 cm long; stipules oblong in basal half, acuminate in upper half, 0.3–0.7 cm long, 0.1–0.3 cm wide, sparsely pubescent, caducous. Inflorescences 2–4, borne on condensed (rarely to 20 cm long), leafless specialized short shoots arising from the axils usually of fallen leaves, 4–19 cm long; peduncle 1.6–9.5 cm long; bracts 0.1–0.3 cm long, 0.03–0.1 cm wide, pubescent, caducous; branches forking 3–4 times; bracteoles not seen; pedicels 0.5–1.8 cm long, tomentose to glabrous; epicalyx bracts inserted on pedicels, linear-lanceolate 0.1–0.2 cm long, caducous. Calyx lobes reflexed, lanceolate, 0.4–0.9 cm long, 0.1–0.3 cm wide, pubescent. Petals spreading, white to pink, 1–1.7 cm long, 0.4–0.8 cm wide. Androecium 0.6–1 cm long; staminal tube dilating towards apex, 0.1 cm long; staminodes 0.6–0.9 cm long; longest stamen 0.6–0.7 cm long, anthers 0.2–0.3 cm long. Ovary globose, tomentose; style 0.1–0.3 cm long, puberulous, 3-branched. Capsule and seeds unknown. Fig. 12: 4, p. 71.

UGANDA. Karamoja District: Taan Valley, *Brasnett* 144!; Teso District: Mbale, Bugwere, Jan. 1931, *Hill* 10!; Busoga District: Kamuli–Buyende road, Nov. 1931, *C. Harris* 446!
KENYA. Uasin Gishu District: Kipkarren, March 1932, *Brodhurst Hill* 685
DISTR. **U** 1–3, **K** 3; Senegal, Gambia, Guinee (Bissau), Guinea, Ghana, Benin, Nigeria, Chad, Cameroon, Central African Republic, Congo-Kinshasa, Sudan, Ethiopia
HAB. Grassland, wooded grassland; 900–1900 m

SYN. *Xeropetalum quinquesetum* Del., Voy. Méroé: 84 (1826)
    X. multiflorum Endl., Nov. Stirp. Decad: 43 (1839).Type: Sudan, Shangul, *Kotschy* 656 (BM, BR (photo.), K, OXF, P, W, syn.), *Kotschy* 685 (W, syn.?)
    X. minus Endl., Nov. Stirp. Decad: 37 (1839). Type: Sudan, 'Shangul et Camamil', *Kotschy* 525 (W, holo., BM, BR (photo.), K, OXF, iso.)
    Dombeya reticulata Mast., F.T.A. 1; 228 (1868); K.Schum. in E.M. 5: 36 (1900). Type: Uganda, Nile Land, *Speke & Grant* 737 pro parte (K, holo.)
    D. alascha K.Schum., E.M. 5: 37 (1900). Type: Ethiopia, Worrhey near Adua, *Schimper* 695 (B†, holo.; BM, K, iso.)
    D. multiflora (Engl.) Planch. var. *vestita* K.Schum. in E.M. 5: 34 (1900) pro parte quoad *Schweinfurth* 2830 & 3264. Type: Congo-Kinshasa, Juru, *Schweinfurth* 3264 (B,†; BR lecto., chosen by Bamps, 1963; K, iso.)
    D. quinqueseta (Delile) Exell var. *vestita* (K.Schum.) Bamps, F.C.B. 10: 250 (1963). Type: Congo-Kinshasa, Juru, *Schweinfurth* 3264 (BR, lecto.)

CONSERVATION NOTES. A widespread and common taxon, here assessed as of Least Concern.

8. **Dombeya rotundifolia** (*Hochst.*) *Planch.*, Fl. Serres 6: 225 (1851); K. Schum. in E.M. 5: 35 (1900); I.T.U.: 146 (1951); T.T.C.L. 598 (1969); Wild in F.Z. 1: 525, t. 98 (1961); Bamps in F.C.B. 10: 248 (1963); Seyani, Dombeya: 173 (1982) & Dombeya in Africa: 96 (1991); K.T.S.L.: 164, map (1994); Vollesen in Fl. Eth. 2, 2: 170 (1995). Type: South Africa, Natal, *Krauss* 252 (W, holo.?; K, BM, RD, OXF, iso.)

Tree to 12 m tall; stems terete or grooved, tomentose to glabrous, hairs simple, stellate and subsessile-glandular; older stems purple-brown. Leaves suborbicular, rarely slightly lobed, 8–21 cm long, 8.5–24 cm wide, apex rounded, rarely acute, base cordate, margin deeply and irregularly dentate; upper surface scabridulous, below tomentellous; petiole 2.5–9 cm long; stipules narrowly lanceolate, 0.2–0.4 cm long, ± 0.1 cm wide, acuminate, pubescent, caducous. Inflorescence resembling that of *D. kirkii* – borne on short specialised shoots, in fascicles of 2–5 in axils of scale-like leaves, each 5–12 cm long, many-flowered, with branches bearing clustered or single flowers; peduncle up to 6 cm long, tomentose; bracts and bracteoles not seen; pedicels 0.6–1.6 cm long; epicalyx bracts inserted at calyx base or pedicel, narrowly elliptic to ligulate, 0.2–0.5 cm long.

Calyx lobes reflexed, lanceolate, 0.5–0.7 cm long, 0.1–0.3 cm wide, acuminate, tomentose. Petals usually spreading, white or pale pink, 0.8–1.4 cm long, 0.5–0.9 cm wide. Androecium 0.6–1 cm long; staminal tube dilating towards apex, 0.05–0.1 cm long; staminodes 0.6–0.9 cm long, longest stamen 0.4–0.7 cm, anthers 0.1–0.2 cm long. Ovary globose, tomentose, style 0.3–0.5 cm long, pilose to glabrous, 3-branched, capsule oblate, 0.5–0.6 cm diameter, setose with a mixture of simple and stellate hairs; seeds ± 0.3 cm long, 0.2 cm wide, light brown. Fig. 11: 7–8, p. 61.

UGANDA. Karamoja District: Warr, Nov. 1939, *Thomas* 3189!; Ankole District: Ruampara, Kigarama Hill, Oct. 1932, *Eggeling* 638!; Mbale District: Elgon, Kaburon, Jan. 1936, *Eggeling* 2487.
KENYA. Northern Frontier District: Mt Nyiru, Tuum–South Horr, Oct. 1978, *Gilbert, Gachathi & Gatheri* 5199!; Turkana District: Lodwar area, Kauwalathe [Kuwalath], Sep. 1963, *Paulo* 1055!; Fort Hall/Nyeri District: 95 km NE Nairobi, Jan. 1972, *Bally & A.R. Smith* B14762!
TANZANIA. Musoma District: Loliondo, Kline's Camp, Nov. 1953, *Tanner* 1747; Kigoma District: Mt Livandabe, 8 May 1997, *Bidgood et al.* 4235!; Mbeya District: Laudan Mwakalinga Forest Reserve, Luanda, Aug. 1965, *Mgaza* 680!
DISTR. U 1–3; **K** 1–4, 6; **T** 1, 2, 4, 7; Congo-Kinshasa, Rwanda, Burundi, Ethiopia, Angola, Zambia, Malawi, Mozambique, Zimbabwe, Botswana, Namibia, Swaziland and South Africa
HAB. Forest edge, woodland, wooded grassland; 1000–2400 m

SYN. *Xeropetalum rotundifolium* Hochst. in Flora 27: 295 (1844)
    Dombeya ringoetii De Wild. in B.J.B.B. 5: 24 (1915). Type: Congo-Kinshasa, Nieuwdorp, *Ringoet* 1 (BR, holo.)
    D. condensiflora De Wild., Contrib. Fl. Kat. Suppl. 1: 46 (1927). Type: Congo-Kinshasa, Haut-Katanga, *Quarré* 574 (BR, holo.; A, iso.)
    D. delevoyi De Wild., Contrib. Fl. Kat. Suppl. 1: 47 (1927). Type: Congo-Kinshasa, riv. Lukuga, *Delevoy* 173 (BR, holo.)
    D. subdichotoma De Wild., Contrib. Fl. Kat. Suppl. 1: 57 (1927). Types: Congo-Kinshasa, ferme de Kibembe, *Quarré* 628 (BR, syn.; A, isosyn.); Lumbumbashi [Elisabethville], *Thomas* 957 (BR, syn.)
    D. spectabilis sensu Mast. in F.T.A. 1: 228 (1868) p.p. quoad *Meller* s.n., *non* Bojer

CONSERVATION NOTES. A widespread and common taxon, here assessed as of Least Concern.
NOTES. Listed in the Mkomazi checklist (**T** 3), but specimen not seen by me.

9. **Dombeya kirkii** *Mast.*, F.T.A. 1: 227 (1868); K. Schum. in E.M. 5: 39 (1900); Wild in F.Z. 1: 527 (1961); Bamps in F.C.B. 10: 247 (1963); Seyani, Dombeya: 152 (1982) & Dombeya in Africa: 112 (1991); Blundell, Wild Fl. E Africa: t. 189 (1987); K.T.S.L.: 164 (1994); Vollesen in Fl. Eth. 2, 2: 170 (1995). Types: Malawi, 16° S, *Meller* s.n. (K, lecto., chosen by Seyani, 1991); Mozambique, Lupata, *Kirk* s.n. (K, syn.)

Small tree or shrub 2–15 m tall; stems terete, densely pubescent to subglabrous, hairs simple, stellate and subsessile-glandular. Leaves ovate to broadly ovate or suborbicular, usually unlobed, rarely shallowly 3-lobed, 3.5–18 cm long, 2–13 cm wide, acuminate or subacuminate, rarely acute or rounded, margin evenly or irregularly serrate or crenate toothed, base cordate, upper surface sparsely stellate-puberulous, lower surface pubescent; petiole 1.5–9 cm long; stipules lanceolate-subulate, 0.4–0.7 cm long, 0.1–0.15 cm wide, pubescent, caducous. Inflorescence axillary, 3–12.5 cm long, the two main branches bearing numerous subumbellate, 3–5-flowered clusters on short branches or single flowers; peduncle 1.9–7.2 cm; bracts and bracteoles not seen; pedicels 0.7–1.5 cm, shortly stellate to villous; epicalyx bracts inserted on calyx base or pedicel, ligulate–oblong, 0.3–0.4 cm long, caducous. Calyx lobes reflexed, lanceolate, 0.4–0.7 cm long, 0.1–0.2 cm wide, acuminate. Petals white, spreading, 0.7–1.3 cm long, 0.4–0.7 cm wide. Androecium 0.6–0.8 cm long; staminal tube dilating towards apex, 0.1 cm long; staminodes 0.5–0.7 cm long; longest stamen 0.3–0.45 cm; anthers ± 0.15 cm long; ovary tube dilating towards apex, 0.1 cm long; staminodes 0.5–0.7 cm long; longest stamen 0.3–0.5 cm; anthers ± 0.15 cm long. Ovary globose, tomentose; style 0.2–0.3 cm long, glabrous, 3-branched. Capsule oblate, 0.4–0.6 cm diameter, tomentose; seeds 0.2–0.3 cm long, ± 0.2 cm wide, dull pale brown, rugose. Fig. 12: 1–3, p. 71.

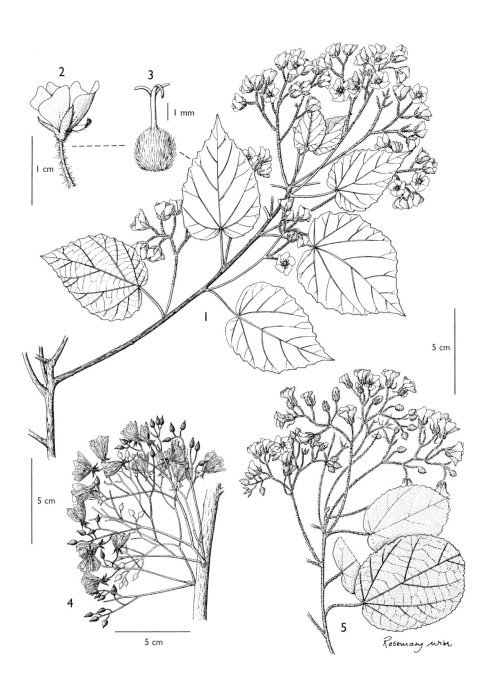

FIG. 12. *DOMBEYA KIRKII* — **1**, habit; **2**, flower; **3**, gynoecium. *DOMBEYA QUINQUESETA* — **4**, leafless flowering stem. *DOMBEYA TAYLORII* — **5**, habit. Drawn by Rosemary Wise. Reproduced from Seyani, Dombeya in Africa, t. 37 and 40 p.p.

UGANDA. Bunyoro District: Bugoma Forest Reserve, Oct. 1962, *Styles* 148!; Busoga District: Jinjaat, Vunamba, Jan. 1932, *C. Harris* 42!; Mengo District: Kampala–Entebbe 14 km, Sep. 1932, *Eggeling* 530

KENYA. Northern Frontier District: Wamba, Nov. 1958, *Newbould* 3151!; Machakos District: Athi R. at Donyo Subuk road, Fourteen Falls, Jan. 1960, *Verdcourt* 2610!; Tsavo Park West: 4 km W Kitani Lodge, Aug. 1965, *Gillett* 16843

TANZANIA. Shinyanga District: Kizumbi, May 1945, *Greenway* 7433!; Moshi District: W end Lake Chala, July 1968, *Bigger* 2000!; Mpanda District: Mahale Mts, Kasoge, May 1995, *Turner* 48!

DISTR. U 2–4; **K** 1–7; **T** 1–5, 7; Congo-Kinshasa, Rwanda, Burundi, Ethiopia, Zambia, Malawi, Mozambique, Zimbabwe, South Africa

HAB. Forest edge, *Acacia* bushland, bushland on rocky slopes; 600–2400 m

SYN. *D. gilgiana* K.Schum. in P.O.A. C: 270, t. 30 (1895); T.T.C.L. 598 (1949). Type: Tanzania, District unclear, Mschusas village, *Holst* 8993 (B†, holo.; BR photo., P, iso.); Usambara-Usagara, Mafumbai, *Holst* 917a (B†, holo.; K, iso.)

  D. gilgiana K.Schum. var. *scaberula* K.Schum., E.M. 5: 39 (1900); T.T.C.L. 598 (1949). Type: Tanzania, Usambara-Usagara, Mafumbai, *Holst* 9171a (B†, holo.; K, iso.)

  D. laxiflora K.Schum. in E.M. 5: 37 (1900). Types: Malawi, *Buchanan* 345 (B, syn†; BR, K, iso.); E Africa, 2°–7° S, *Hannington* s.n. (K, syn.)

  D. umbraculifera K.Schum. in E.M. 5: 38 (1900). Type: Kenya, Masai, Kitui, Ukamba, *Hildebrandt* 2780 (B†, holo.)

  D. mukole Sprague in J.L.S. 37: 502 (1906); I.T.U.: 419 (1951); Hamilton, Uganda For. Trees: 118 (1981). Types: Uganda, Mengo District: Mabira Forest, *Dawe* 182, 394; Bunyoro District: Budongo Forest, *Dawe* 831 (all K, syn.)

  D. warneckei Engl. in E.J. 39: 584 (1907) & V.E. 3(2): 431, t. 200 (1921); T.T.C.L.: 598 (1949). Types: Tanzania, Lushoto District: Sigi below Amani, *Warnecke* 480 (A, E, EA, K, P, syn.); Dodoma District: Kilimatinde, Mukundaku–Konko road, *von Prittwitz* 171 (B†, syn.)

  D. kituiensis Hochr. in Candollea 3: 54 (1926). Type: Kenya, Kitui in Ukamba, *Hildebrandt* 2780 (P?, holo.; BR, phot.)

  D. gillettii Gilli in Ann. Naturhist. Mus. Wien 74: 448 (1971). Type: Kenya, Northern Frontier District: Dandu, *Gillett* 13103 (W, holo.; BR, EA, K, iso.)

CONSERVATION NOTES. A widespread and common taxon, here assessed as of Least Concern.

10. **Dombeya taylorii** *Baker f.* in J.B. 39: 125 (1901); Seyani, Dombeya: 190 (1982) & Dombeya in Africa: 117 (1991); K.T.S.L.: 164 (1994); Thulin, Fl. Somal. 2: 21, fig. 9 (1999). Type: Kenya, Mombasa Island, *Taylor* s.n. (BM, holo.)

Small tree or shrub to 6 m tall resembling *D. kirkii*, but leaves obovate or at least widest in upper half, not exceeding 8 cm long, apex rounded, venation below prominent, reticulate; inflorescence indumentum densely villous, hairs up to 1.5 mm long; epicalyx bracts filiform; petals shorter, 0.5–0.9 cm long. Fig. 12: 5, p. 71.

KENYA. Kilifi District: Kaya Fimboni, Aug. 1989, *Robertson & Luke* 5826!; Mombasa District: Mombasa–Lamu, Jan. 1903, *Whyte* s.n.!; Kwale District: Mrima Hill, Sep. 1957, *Verdcourt* 1901!

TANZANIA. Lushuto District: 1 mile E Mashewa, June 1953, *Drummond & Hemsley* 3077!; Handeni District: Kwa Mkono, July 1983, *Archbold* 2986!; Pangani District: Pangani, Mufu, Msero, July 1955, *Tanner* 1934!

DISTR. **K** 7; **T** 3; Somalia

HAB. Thicket or wooded grassland, rarely edge of lowland evergreen forest; 0–500(–1000) m

SYN. *D. kirkii* sensu K.Schum. in E.M. 5: 39 (1900) p.p. quoad *Holst* 3180 & *Wakefield* s.n., *non* Mast.

  D. praetermissa Dunkley in K.B. 1934: 182 (1934); T.T.C.L.: 598 (1949). Type: Kenya, Kilifi Coast, *Graham* 350 (K, holo.; A, BR, FHO, iso.)

CONSERVATION NOTES. This species is largely confined to the coastal thicket on each side of the Kenyan-Tanzanian border; 24 specimens are known, derived from four districts in each country. One collection is known from Somalia. While there is pressure on coastal forest patches, there is little commercial value, so little threat, to coastal thicket in this part of E Africa (Verdcourt pers. comm. 2005). Accordingly *D. taylorii* is here assessed as near threatened.

9. MELHANIA

Forssk., Fl. Aegypt.-Arab.: cvii, 64 (1775); Bayer & Kubitzki in Fam. Gen. Vasc. Pls. 5: 268 (2003)

Brotera Cav., Icon. 5: 19 (1799), *non* Spreng. (1801), *nec* Spreng. (1802), *nec* Willd. (1800), *nec* Vell. (1825)
Sprengelia Schult., Observ. Bot.: 134 (1809), *non* Sm. (1794)

Suffrutescent annual or perennial herbs, subshrubs, or shrubs; erect, ascending or decumbent; young stems tomentose, the hairs varied, mostly stellate, sparsely pubescent in age. Leaves simple, petiolate, stipulate. Flowers bisexual, rarely polygamous. Inflorescences 2–4-flowered, axillary (FTEA) or terminal cymes, occasionally appearing subumbellate or racemose (not in FTEA), or flowers solitary; peduncle and pedicels tomentose; epicalyx bracts 3, close to the calyx, sometimes fused at base, persistent, either linear to ovate or broadly ovate and not accrescent nor becoming membranous in fruit or obovate, cordate to reniform and enlarging and becoming membranous in fruit. Floral buds with sepal tips either free or confluent. Sepals 5, almost free, with a narrow patch of glandular tissue at the base of each lobe, persistent, sometimes accrescent. Petals symmetrical to strongly asymmetrical, yellow, usually opening in the afternoon, caducous or marcescent, sometimes persistent with fruit after falling. Stamens and staminodes united into a very short staminal tube, stamens 5, alternating with 5 ligulate staminodes. Ovary syncarpous, 5-locular, densely stellate-pubescent, 1–12 (or more?) ovules per locule; style apically divided or lobed; stigmatic lobes 5, slender. Capsule spheroid to ovoid, hardened to chartaceous, pubescent, loculicidally dehiscent; endocarp glabrous or pubescent; 1–many seeds per locule. Seeds ± trigonal to turbinate, 3-(more)-angled, testa ± smooth, sparsely to densely tuberculate, or muricate, elaisome present or absent; endosperm abundant; cotyledons folded and bipartite.

A genus of about 50–60 species in tropical and subtropical areas of Africa, Arabia, Madagascar, Asia, and Australia; centres of species diversity appear to be in northeastern Africa (some 25 species) and in southwestern Madagascar (some 18 species).

Schumann (1900) divided the African species of *Melhania* into three subgenera based on epicalyx bract characters. These were *M.* subgen. *Broteroa* K.Schum., with oblong-lanceolate to lanceolate epicalyx bracts that are neither accrescent nor membranous; subgen. *Melhania* (as *Eumelhania* K.Schum.), with ovate or broadly ovate epicalyx bracts that are neither accrescent nor membranous; and subgen. *Hymenonephros* K.Schum. with reniform to widely cordiform epicalyx bracts that are both accrescent and membranous in fruit. The distinction between the first two subgenera appears to be artificial and Arènes (in Fl. Madag. 131: 160 (1959)) may have perceived this when he reduced them to sections. More than shape, texture and whether or not epicalyx bracts are accrescent in fruit seem to be correlated characters that distinguish two large groups of *Melhania* species.

1. Epicalyx bracts longer than wide, subulate to
    ovate, not or slightly accrescent but never
    membranous enlarged and membranous in
    fruit 2
   Epicalyx bracts wider than long, or equally as
    wide as long, broadly ovate, widely cordate, or
    reniform, enlarged and membranous in fruit 11
2. Epicalyx bracts ovate to broadly ovate 3
   Epicalyx bracts linear to narrowly ovate 6
3. Annual plants, to 15 (–30) cm tall; leaves ovate
    to obovate; capsules longer than wide (4–8 ×
    3–5 mm); 2–4 seeds per locule . . . .   3. *M. annua* (p. 77)
   Annual or perennial plants, to 1 m tall or taller;
    leaves narrowly ovate, ovate or broadly ovate
    (rarely round or obovate); capsules as long as
    wide (6–14 × 6–12 mm); 4–12 seeds per locule 4

4. Largest leaves broadly ovate to round, length:
   breadth 1–1.5:1 .... 7. *M. rotundata* (p. 83)
   Largest leaves ovate to narrowly lanceolate,
   length; breadth 2–6:1 5
5. Leaves less than 3 times as long as wide; epicalyx
   bracts abruptly acuminate, conspicuously
   wider at the base .... 1. *M. velutina* (p. 75)
   Leaves usually 3 or more times as long as wide;
   epicalyx bracts acute to acuminate, but never
   abruptly so, not noticeably wider at the base . 2. *M. angustifolia* (p. 76)
6. Leaf margin entire (rarely shallowly dentate in
   part); leaves and inflorescences arising from a
   stout caudex .... 8. *M. randii* (p. 83)
   Leaf margin serrate to dentate or crenulate;
   leaves and inflorescences not arising from a
   stout caudex 7
7. Epicalyx bracts linear to subulate or narrowly
   lanceolate, often narrower than the sepals 8
   Epicalyx bracts lanceolate to narrowly ovate,
   usually as wide or wider than the sepals 9
8. Stems whitish lanate; stipules (0.8–)1–2 cm long 9. *M. volleseniana* (p. 84)
   Stems whitish tomentose, not lanate; stipules
   0.2–0.9 cm long .... 5. *M. ovata* (p. 80)
9. Leaves narrowly ovate, narrowly elliptic, or oblong
   (rarely lanceolate); epicalyx bracts not usually
   reflexed; apices of sepal lobes very briefly free
   in bud; style (1–)2–3 mm long; seed coat
   densely muricate .... 6. *M. parviflora** (p. 81)
   Leaves broadly ovate or widely elliptic to round;
   epicalyx bracts usually reflexed (tardily in
   M.denhardtii); apices of sepal lobes free in
   bud; style 4–6 mm long; seed coat sparsely
   tuberculate to smooth or striate and then
   appearing smooth to the naked eye 10
10. Epicalyx bracts 7–16 (20) × 2.5–5 (7) mm;
    capsules ellipsoid, flattened apically; 4–6 seeds
    per locule; seed coat sparsely tuberculate to
    smooth .... 7. *M. rotundata* (p. 82)
    Epicalyx bracts 5–7 × 1–2 mm; capsules ovoid to
    narrowly ovoid, not flattened apically; 3 seeds
    per locule; seed coat striate and appearing ±
    smooth to the naked eye .... 4. *M. denhardtii* (p. 78)
11. Flowers solitary (rarely in 2-flowered cymes);
    epicalyx bracts cordiform with a deep basal
    sinus equalling $\frac{1}{4}$ to $\frac{1}{3}$ the height of the bract
    in fruit, basal lobes of bracts sometimes
    overlapping .... 12. *M. substricta* (p. 87)
    Flowers in 2–8-flowered cymes; epicalyx bracts
    widely cordiform to reniform with a shallow
    basal sinus equalling $\frac{1}{4}$ or less the height of
    the bract in fruit, basal lobes of bract never
    overlapping 12

12. Shrubby perennials to 2 m tall; stems ± yellowish
　　 tomentose; petals 7–10 mm long; capsules
　　 7–10 mm long; 2–3 seeds per locule; seed coat
　　 tuberculate ....　10. *M. phillipsiae* (p. 86)
　　 Herbaceous perennials or subshrubs to 0.6 m tall;
　　 stems lanate, whitish tomentose, or canescent;
　　 petals 3.5–5 mm long; capsules 3–5 mm long; 1
　　 seed per locule; seed coat smooth 13
13. Leaves 3 times as long as wide, margin entire or
　　 denticulate only in the upper half, larger
　　 leaves never lobed; young stems canescent;
　　 seeds with an umbonate projection ....　13. *M. praemorsa* (p. 88)
　　 Leaves less than 3 times as long as wide, margin
　　 shallowly and distantly crenate to dentate,
　　 larger leaves often shallowly lobed; young
　　 stems lanate, whitish tomentose; seeds without
　　 an umbonate projection ....　11. *M. latibracteolata* (p. 86)

* Species 14, *M. polyneura*, which is imperfectly known, would probably key out as *M. parviflora*.

1. **Melhania velutina** *Forssk.*, Fl. Aegypt.-Arab.: cvii, 64 (1775); E.P.A.: 579 (1959); Bamps in F.C.B. 10: 253 (1963); F.P.U.: 58 (1971); U.K.W.F.: 191, fig. p. 192 (1974); Vollesen in Opera Bot. 59: 34 (1980); Blundell, Wild Flow. Kenya: 47 (1982); Troupin, Fl. Rwanda 2: 403 (1983); Blundell, Wild Flow. E. Afr.: 73 (1987); Sapieha, Wayside Flow. E. Afr.: t. 287 (1987); U.K.W.F., ed. 2: 97 (1994); Vollesen in Fl. Eth. 2(2): 174–175, figs. 80.4/4, 80.5/1–4 (1995); Thulin, Fl. Som. 2: 26, fig. 12, A-D (1999). Types: Yemen, Jebel Milhan [in monte Melhân], 1763, *Forsskål* s.n. (BM!, syn.); *Forsskål* 510 (C, syn.); *Forsskål* 1909 (LIV, syn.)

Perennial herb or subshrub, to 1(–2.5) m tall; erect; young stems pale yellowish to brownish or ferruginous pubescent to tomentose or velutinous, the hairs sessile-stellate, stalked-stellate, and tufted; sparsely pubescent in age. Leaves ovate to broadly ovate (rarely some leaves obovate), 2.5–13 cm long, 1–6 cm wide, apex rounded to emarginate, margin shallowly serrate, base rounded (rarely somewhat cuneate), tomentose above, the hairs stellate or long-armed tufted, paler below and usually more densely stellate, especially along veins and leaf margin, the hairs stellate or long-armed tufted and often reddish-brown, especially along the veins and leaf margin; palmately 3–5-nerved from the base; petiole 0.6–3 cm long, ± 1–2 mm wide, reddish-brown, golden-brown to light-brown velutinous; stipules filiform (rarely subulate), 0.4–1.7 cm long, reddish brown when dry (with scattered stellate hairs), tardily dehiscent. Inflorescences 2–4-flowered cymes, occasionally appearing subumbellate, or flowers solitary; peduncle 1.8–8.5 cm long; pedicels 0.2–4 cm long; both peduncle and pedicels velutinous to tomentose; epicalyx bracts broadly ovate, asymmetrical, 4–12 mm long, 2–8 mm wide, often slightly accrescent in fruit, apex sharply narrowing, acuminate, base cuneate to very shallowly cordate, reddish-brown, golden-brown to pale-yellow velutinous. Floral buds widely ovoid, the sepal tips free. Sepals narrowly ovate, 7–12 mm long, 2–3 mm wide, often slightly accrescent in fruit, velutinous without, glabrous within. Petals obovate, asymmetrical, 6–10 mm long, 4–7 mm wide, yellow, glabrous. Staminal tube ± 1 mm long, free portion of filaments 1–1.5 mm long, anthers ± 1.5 mm long, free portion of staminodes 4–5 mm long, glabrous. Ovary ovoid, 2–3 mm long, 2–3(–7) mm in diameter, densely velutinous; 10–12 ovules per locule; style 0.5–2.5 mm long, briefly pubescent at base; stigma lobes 0.5–1 mm long, erect to spreading or recurved. Capsule ovoid, sometimes pointed at the apex, 6–10 mm long, 6–9 mm in diameter, tomentose; 4–12 seeds per locule; seed trigonous, ± 1.7 mm long, ± 1.2 mm wide, testa scatteredly tuberculate, without an elaisome.

UGANDA. Karamoja District: Lomaler–Kakamari road, June 1930, *Liebenberg* 218!; Toro/Ankole District: Lake George flats, Sep. 1938, *Purseglove* 397!; Teso District: Serere, Dec. 1931, *Chandler* 100!

KENYA. Machakos District: Kibwezi, Mar. 1921, *Dummer* 4603!; Kisumu-Londiani District: Nyahera Hills–Kisumu, 17 Apr. 1965, *Kokwaro* 61!; Kilifi District: Jilore, Arabuko-Sokoke Forest Reserve, 28 Nov. 1961, *Polhill & Paulo* 874!

TANZANIA. Mbulu District: Tarangire National Park, road to Burungi, 12 Feb. 1970, *Richards* 25407!; Lushoto District: Lushoto–Mombo road, 2 km SW of Gare turnoff, 16 June 1953, *Drummond & Hemsley* 2929!; Uzaramo District: Oyster Bay, 26 June 1968, *Batty* 172!; Zanzibar I., 1931, *Vaughan* 1066!

DISTR. U 1–4; **K** 3?,4–7; **T** 1–6, 8; **Z**; Congo-Kinshasa, Rwanda, Burundi, Sudan, Eritrea, Ethiopia, Somalia, Malawi, and possibly Angola; also in Saudi Arabia and Yemen

HAB. *Acacia - Commiphora, Acacia - Combretum - Terminalia*, and *Boswellia* woodland, wooded grassland and bushland, riverine forest, lowland and medium altitude woodland; also in cultivated areas; (sea-level–)700–1950 (–2250) m

SYN. *Pentapetes velutina* (Forssk.) Vahl, Symb. Bot. 1: 49 (1790)
    Melhania ferruginea A.Rich., Tent. Fl. Abyss. 1: 76 (1847); Mast. in F.T.A. 1: 231 (1868); K. Schum. in Engl., Hochgebirgsfl. Trop. Afr.: 303 (1892); P.O.A. C: 269 (1895); Baker f. in J.B. 36: 5 (1898); K. Schum. in E.M. 5: 14–15, t. 1, fig. C (1900); De Wild., Pl. Bequaert. 1: 515 (1922); Cufod. in Miss. Biol. Borana, Racc. Bot. 4: 135 (1939); F.P.N.A. 1: 604 (1948); T.T.C.L.: 601 (1949); F.P.S. 2: 5–6 (1952). Type: Ethiopia, "Aderbati," Taccazé valley, Sep., *Quartin-Dillon & Petit* 38 (P, holo.; K!, iso.)
    M. malacochlamys K.Schum. in E.M. 5: 13–14 (1900); T.T.C.L.: 601 (1949). Type: Tanzania, Mwanza District: Kagehi, Dec. 1885, *Fischer* 79 (B†, holo.)

LOCAL USES. In Kenya, string is made from the fibres of the stems (*Glover et al.* 672).

CONSERVATION NOTES. This species appears to be widespread and common, and its habitat not significantly threatened as far as is known. It is rated here as of "least concern" for conservation.

NOTE. The stellate hairs with reddish-brown arms are a distinctive, if somewhat inconstant, feature of *M. velutina*. Almost all the other East African species possess stellate hairs with reddish-brown stalks or reddish-brown lepidote hairs, but the arms of these are never pigmented. *Melhania forbesii* Mast. of southern Africa is very similar to *M. velutina* (it apparently differs in having larger epicalyx bracts and generally more oblong leaves). Schumann (1900) also considered the number of ovules in each cell to be a key difference; 6 in *M. forbesii* versus 10–12 in *M. velutina*. This, however, does not seem to be a reliable distinction since a cursory examination of several specimens of *M. forbesii* from Botswana show there to be as many as 9 seeds per locule. *Melhania forbesii* may prove after further study to be a synonym of *M. velutina*.

According to Thulin (1999: 26), *M. engleriana* K.Schum. is close to *M. velutina* but if the illustration provided by Schumann (in E.M. 5: t. 1, fig. F, a & b.1900) is correct the relationship would seem to be with *M. ovata* (Cav.) Spreng. and relatives.

There is one chromosome number report (n=30) for *M. velutina* vouchered by a collection from Kenya (Bates in Gentes Herb. 10: 39–46, 1967).

2. **Melhania angustifolia** *K.Schum.* in E.M. 5: 11–12 (1900). Type: Tanzania, Zanzibar, *Stuhlmann* I 716 (B†, holo.; BM!, sketch of type)

Suffruticose perennial herb or shrub to 2 m tall; erect, with ± virgate branches; young stems ferruginous or greyish velutinous, minutely stellate-pubescent to glabrescent in age. Leaves narrowly ovate to narrowly oblong, 2.9–8.3 cm long, 0.9–1.9 cm wide, apex acute, margin entire to very shallowly serrulate, base rounded, often oblique, velutinous above and below, palmately 3–5-nerved from the base; petiole 0.6–2 cm long, velutinous; stipules filiform, 0.5–1.5 cm long, ± caducous, sometimes persisting. Inflorescences 2–3-flowered cymes; peduncle 1.2–4.2 cm long; pedicels 0.3–1.1 cm long; peduncle and pedicels velutinous; epicalyx bracts widely triangular, asymmetrical, 8–14 mm long, 5–11 mm wide, enlarged and membranous in fruit, apex acute to acuminate, slightly recurved, base obtuse to truncate, sides reflexed, velutinous. Floral buds hidden by the epicalyx bracts, widely ovoid, the sepal tips briefly free. Sepals narrowly ovate, 6–11 mm long, 2–4 mm wide, tomentose without, glabrous within. Petals broadly ovate, asymmetrical, truncate apically, 8–11 mm long, 5–7 mm wide, bright yellow, glabrous. Staminal tube 1–1.5 mm long, free portion of filaments 1–1.5 mm long, anthers ± 2 mm long, free portion of staminodes 5–7 mm long, glabrous. Ovary widely ovoid, 2–4 mm long, 3–4 mm in diameter, stellate-pubescent; 4–6 ovules per locule; style 3.5–5 mm long; stigma lobes ± 1–1.5 mm long, ascending. Capsules globose to ovoid, 7–10 mm long, 7–10 mm in diameter, 4–6(–10?) seeds per locule; seed ± trigonal to turbinate, ± 1.5 mm long, ± 1.5 mm wide, testa very finely verrucose, but smooth to the naked eye, without an elaiosome.

TANZANIA. Zanzibar, Chukwani, 1 Sep. 1959, *Faulkner* 2344! & Mkunduchi, 27 Nov. 1930, *Greenway* 2590! & Chwaka road, 18 July 1950, *R.O. Williams* 49!
DISTR. **Z**; endemic
HAB. Thick bushland on coral rag or rock, also on white sand; 0–5 m

CONSERVATION NOTES. This species appears to be known from only seven specimens on Zanzibar, and probably fewer sites, most of which are bush on coastal coral rag now threatened by hotel building and other tourist-linked development. Accordingly it is here assessed as Vulnerable (VU B2a, b(iii)). However, site studies may show that it may be far from the brink of extinction. One specimen states "dominant in abandoned native cultivations on coral".
NOTES. The flowers are said to open in the late afternoon (*Faulkner* 2344).
    Blundell (Wild. Flow. E. Afr.: 72–73, t. 287. 1987) included *M. angustifolia* in his guide and reported that the plant illustrated had been photographed in the Taita Hills. I am reluctant to accept the presence of *M. angustifolia* in Kenya on the basis of this photograph since it fails to show critical epicalyx characters. Blundell's report evidently was not vouchered (apart from the photograph) and it does seem, given that the species is otherwise known only from Zanzibar, a bit improbable. The taxon is reported in the Mkomazi checklist (**T** 3), but I have not been able to verify the record.

3. **Melhania annua** *Thulin* in Nordic J. Bot. 19: 193, fig. 1 (1999) & Fl. Som. 2: 26–27, fig. 13 (1999). Type: Somalia, Galguduud: 18 km on road from Masagaweyn to Bud Bud, *Thulin, Hedrén & Abdi Dahir* 7414 (UPS, holo.; K!, iso.)

Annual herb, to ± 15(–30) cm tall; erect or decumbent; young stems pubescent with sessile, stellate, and tufted hairs, becoming sparsely pubescent in age with predominantly sessile-stellate hairs with appressed arms. Leaves broadly ovate to obovate, 0.9–6.2 cm long, 0.5–4.2 cm wide, apex retuse to rounded, mucronate, margin serrate to crenate, base cordate to truncate, slightly discolorous, paler below, pubescent above with mostly sessile-stellate hairs, and below with sessile-stellate hairs; palmately (3–)5-nerved from the base; petiole 0.3–4 cm long, stellate-pubescent; stipules filiform to subulate, (1–)2–8 mm long, reddish brown when dry with very few scattered stellate, bifurcate, and simple hairs, caducous. Inflorescences 2–3-flowered cymes or flowers solitary; peduncle (0.3–)0.5–3 cm long, ± 0.5 mm wide; pedicels 2–10 mm long; peduncle and pedicels stellate pubescent; epicalyx bracts reddish, broadly ovate to cordiform, (2–)4–8 mm long, (1–)2–6(–8) mm wide (in flower), slightly accrescent, (3–)6–13 mm long, (1.5–)4–10 mm wide (in fruit), apex acute or slightly acuminate, base cordate or rarely cuneate, subsessile to short stipitate, sparsely stellate-pubescent. Floral buds obscured by the epicalyx bracts. Sepals narrowly-lanceolate to lanceolate, (2.5–)4–10 mm long, 1–2.5 mm wide, accrescent in fruit, stellate pubescent without, glabrous within. Petals broadly obovate, asymmetrical, ± (2.5–)8 mm long, (1–)3–4(–5.5) mm wide, yellow, glabrous. Staminal tube 0.5–2 mm long, free portion of filaments ± 1 mm long, anthers ± 1 mm long, free portion of staminodes 2.5–5 mm long, glabrous. Ovary globose, 1–2 mm long, 1–2 mm in diameter, densely velutinous; 4 ovules per locule; style 2–4 mm long, glabrous; stigma lobes 1–1.5 mm long, spreading. Capsules subglobose to ovoid-oblong, somewhat flattened at the apex, 4–8 mm long, 3–5 mm in diameter, stellate pubescent, 2–4 seeds per locule; seed ovoid-trigonous to turbinate, 1.5–2.3 mm long, 1–1.5 mm wide, densely tuberculate, with a grey surface and a small, smooth, and darker-colored apical process, elaiosome not present. Fig. 13, p. 79.

KENYA. Lamu District: Kitwa Pembe Hill and vicinity, 0–50 m, 15–16 July 1974, *Faden & Faden* 74/1107! & Kiunga Archipelago, 28 July 1961, *Gillespie* 75!
DISTR. **K** 7; coastal Somalia
HAB. Sand dunes and eroded sand dunes in grey-brown sand along the coast, locally common; 0–50 m.

CONSERVATION NOTES This species is restricted to a stretch of coastal dunes 1000 km long (protologue) and up to 20 km wide (Willdenowia 18, 1989). Nine specimen-sites are known from Somalia and three from Kenya. No information is available on whether there are direct threats to the habitat. Accordingly *M. annua* is here assessed as near threatened.
NOTES. Thulin (1999) allied this species with the widespread *M. ovata* (Cav.) Spreng. and the East and North East African *M. parviflora* Chiov., but nonetheless thought its affinities were obscure. I rather suspect that its affinities lie instead with *M. velutina* and *M. angustifolia.* The epicalyx bract shape, sepal shape, stigmatic surfaces, and capsule shape of *M. annua* all argue for this relationship (and against one with *M. ovata*).

The species appears to be restricted to the "*Acacia* woodland or bushland on coastal dunes" described by Friis and Vollesen (Willdenowia 18: 455–477 (1989)), who speculated that a considerable portion of the sand present in these dunes that range from northeastern Somalia to northeastern Kenya is derived from coral. These dunes evidently harbour many endemic or near endemic species.

4. **Melhania denhardtii** *K.Schum.* in N.B.G.B. 2: 302 (1899), as *dehnhardtii* & in E.M. 5: 8, t.1/E (1900). Type: Kenya, Tana River District: Ngao [Ngad], *Thomas* 139 (B†, holo., BM!, K!, iso.)*

Suffrutescent perennial herb or subshrub, 0.4–1 m tall, erect; young stems reddish, sparsely stellate-pubescent, glabrous in age. Leaves widely elliptic to broadly ovate or obovate, very slightly asymmetrical, 0.7–3 cm long, 0.5–2 cm wide, apex retuse to shortly acute, margin serrate to dentate, base subcordate to truncate, very short tomentose above and below, palmately (3–)5-nerved from the base; petiole 0.3–1.3 cm long, canescent; stipules filiform, 2–3 mm long, caducous. Inflorescences 2–3-flowered cymes or flowers solitary; peduncle 0.9–2.5 cm long; pedicels 5–13 mm long; peduncle and pedicels ± appressed stellate; epicalyx bracts lanceolate to narrowly ovate, 5–7 mm long, 1–2 mm wide, stellate-pubescent often with two class sizes of stellate hairs, tardily reflexed. Floral buds ovoid, the darkened sepal apices free. Sepals narrowly ovate, 7–9 mm long, 1.5–3 mm wide, long acuminate, tomentose without, glabrous within. Petals broadly obovate, asymmetrical, 7–12 mm long, 4–8 mm wide, bright yellow, glabrous. Staminal tube 0.5–1 mm long, free portion of filaments 1–2 mm long, anthers 1–2.5 mm long, free portion of staminodes 4–7 mm long, glabrous. Ovary ovoid, 2–2.5 mm long, 1–2 mm in diameter, tomentose; 3 ovules per locule; style 4–6 mm long, stigma lobes 1–2 mm long, recurved. Capsules ovoid to narrowly ovoid, 4–6 mm long, 5–6 mm in diameter, velutinous, 3 seeds per locule; seed trigonal, 1.5–2 mm long, 1–1.5 mm wide, testa sparsely striate (almost smooth to the naked eye), without an elaiosome.

KENYA. Kilifi District: Mida, Oct. 1930, *Donald* 2499! & Kilifi North, no date, *Graham* 1641! & Dakawachu, 29 Nov. 1990, *Luke & Robertson* 2524!
DISTR. **K** 7; endemic, but expected in neighbouring Somalia
HAB. Open *Acacia-Commiphora* bushland on sandy or rocky soil; 5–250 m

CONSERVATION NOTES. This species is rated here as VU D2, that is vulnerable to extinction by stochastic changes being known from less than 5 sites (all in Kilifi District) with an area of occupancy of less than 20,000 km².
NOTE. *Melhania denhardtii* is most closely related to *M. rotundifolia*, from which it differs by its smaller leaves, smaller epicalyx bracts, fruit shape, and seed surface.

The specific epithet was originally spelled "*dehnhardtii*". It undoubtedly commemorates Clemens (1852–1928) and/or Gustav Denhardt (1856–1917), who briefly formed a German colony at Witu in the late 19th century, and who organized the Tana expedition for which F. Thomas collected plants. The spelling of the specific epithet therefore is correctable under Art. 60.1 of the ICBN.

* The isotypes are labelled "Witu", which is downstream from Ngao

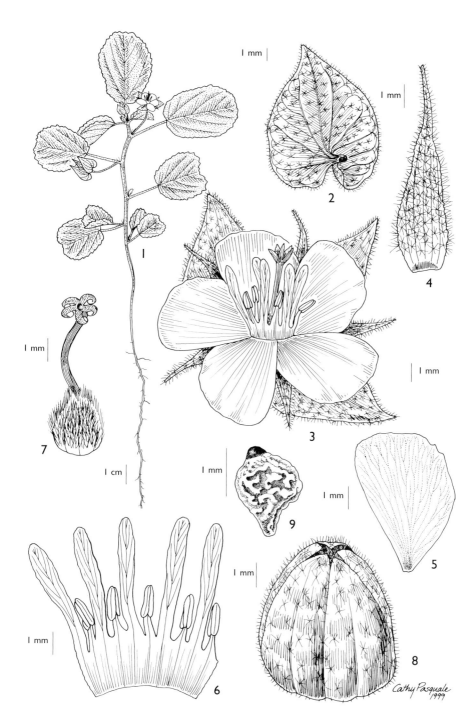

FIG. 13. *MELHANIA ANNUA* — **1**, habit; **2**, epicalyx bract; **3**, flower; **4**, sepal; **5**, petal; **6**, androecium, showing alternating stamens and staminodes; **7**, gynoecium, showing ovary, style and stigmas; **8**, capsule; **9**, seed. 1,2, 5 from *Faden & Faden* 74/1107; 2, 4, 6?9, from *Gillespie* 75. Drawn by Cathy Pasquale.

5. **Melhania ovata** (*Cav.*) *Spreng.*, Syst. Veg. 3: 32 (1826); K. Schum. in Engl., Hochgebirgsfl. Trop. Afr.: 303 (1892) & in P.O.A. C: 269 (1895) & in E.M. 5: 6–7, t. 1, fig. D (1900); F.W.T.A., ed. 2, 2(1): 318 (1958); U.K.W.F.: 191 (1974); Blundell, Wild Flow. E. Afr.: t. 288 (1987); U.K.W.F., ed. 2: 97 (1994); Vollesen in Fl. Eth. 2(2): 173, fig. 80.4/1 (1995); Thulin, Fl. Som. 2: 24 (1999). Type: Cult. "R. H. Matritense"[Madrid Bot. Gard.] ex seed from "prope Huanajuato in Nova-Hispania," Sep. 1798, (MA, holo.) Although Cavanilles stated that the provenance of the seed for this specimen was Mexico (i.e., near Guanajuato), the species is confined to the Old World tropics and the seed probably originated in India

Shrubby herb or suffrutescent subshrub, to 1 m tall (stems short and densely-branched at elevations above 1300 m in FTEA area); young stems canescent or greenish-tomentose, the hairs stellate or tufted, the arms ± appressed and tending to lie in the same direction, also with stellate hairs with reddish-brown centres on youngest parts, ± glabrescent in age. Leaves ovate to elliptic, 0.5–8.5 cm long, 0.6–4.4 cm wide (usually 0.5–1.2 cm long, 0.6–1 cm wide or smaller at elevations above 1300 m in FTEA area), apex rounded to emarginate, margin irregularly serrate to dentate, the larger leaves ± crenate, base rounded to somewhat cuneate, tomentose above, the hairs stellate, paler below and more densely stellate, with scattered stellate hairs with reddish-brown centres; palmately 3–5-nerved from the base; petiole 0.2–1.5 cm long, tomentose, the hairs long-armed stellate, with scattered stellate hairs with reddish-brown centres; stipules filiform to subulate, 2–9 mm long, often persisting. Inflorescences 2–3-flowered cymes, or flowers solitary; peduncle 0.5–4.5 cm long; pedicels 0.3–1.4 cm long (or sometimes smaller in plants from elevations above 1300 m in FTEA); both peduncle and pedicels tomentose when young, becoming ± glabrous in age; epicalyx bracts subulate to filiform, 3–13 mm long, ± 1 mm wide, greyish-green tomentose. Floral buds narrowly ovoid, the sepal tips confluent. Sepals narrowly ovate, 5–14 mm long, 2–3 mm wide, sometimes long apiculate, tomentose without, glabrous within except for tomentose apex. Petals obovate, slightly asymmetrical, 4–8 mm long, 4–6 mm wide, yellow, glabrous. Staminal tube 1–2 mm long, free portion of filaments 1–2 mm long, anthers 1.5–2 mm long, free portion of staminodes 3–6 mm long, glabrous. Ovary ovoid, 2–3 mm long, 2–3 mm in diameter, velutinous; 6 ovules per locule; style 0.5–2(–3) mm long; stigma lobes 0.5–1(–2) mm long, recurved. Capsules ovoid, 5–9 mm long, 5–8 mm in diameter, often with a slight apical rostrum, tomentose, 3–4(–5) seeds per locule; seed trigonal, with two flattened and one rounded side, ± 2 mm long, ± 1.5 mm wide, testa moderately tuberculate, without an elaiosome.

UGANDA. Bunyoro District: Bugungu Hill, 6 Jan. 1922, *Dummer* 5520!
KENYA. Northern Frontier District: Furrole, 14 Sep. 1952, *Gillett* 13873!; Laikipia District: 30 km N of Rumuruti, 7 Nov. 1978, *Hepper & Jaeger* 6648!; Nairobi National Park, Feb. 1963, *Williams Sangai* 775!
TANZANIA. Musoma District: Serengeti National Park, Naabi Entrance, 26 May 1962, *Greenway* 10331!; Masai/Mbulu District: Ngorongoro crater floor, East, Apr. 1941, *Bally* B 2410! & Ngorongoro Conservation Area: Olduvai, 2 June 1977, *Raynal* 19358!
DISTR. **U** 2; **K** 1, 3, 4, 6, 7; **T** 1–3, 5; Cape Verde Islands, Senegal, Mauritania, southern Sudan (fide Blundell), Eritrea, Ethiopia, and Somalia; also in Arabia, India, and Australia
HAB. *Acacia*, *Acacia-Combretum*, and *Acacia-Balanites* wooded grassland and bushland, mostly on stony soil and rocky slopes, often in overgrazed areas; 500–1900 m

SYN. *Brotera ovata* Cav., Icon. 5: 20, t. 433 (1799); McVaugh, Bot. Res. Sessé & Mociño Exped. 7: 518 (2000)
　　Pentapetes ovata (Cav.) Willd., Enum. Pl.: 719 (1809)
　　?Melochia ovata Desf., Tabl. École Bot., ed. 2, 172 (1815)
　　Melhania abyssinica A.Rich., Tent. Fl. Abyss. 1: 76 (1847) & Tent. Fl. Abyss., Atlas: t. 18 (1851?); Mast. in F.T.A. 1: 231–232 (1868, but as "*M. ovata*" in the key!); W.F.K.: 34 (1948) ("*Melhamia*"). Type: Ethiopia, Aderbati, Taccazé valley, Sep., *Quartin-Dillon & Petit* s.n. (P, holo.)
　　M. oblongata Mast. in F.T.A. 1: 232. 1868, *nom. nud.*, *pro syn.* Material cited: Ethiopia, near Dschadscha, *Schimper* 2108 (K!, S)

M. ovata (Cav.) Spreng var. *abyssinica* (A.Rich.) K.Schum. in E.M. 5: 7–8 (1900)
M. ovata (Cav.) Spreng var. *oblongata* K.Schum. in E.M. 5: 7 (1900). Types: Kenya, near
  Mombasa, *Kirk* s.n. (K, syn.) [and many additional syntypes from elsewhere in Africa; a
  number of which have been referred to other taxa of *Melhania*]
M. ovata (Cav.) Spreng var. *montana* K.Schum. in E.M. 5: 8 (1900); E.P.A.: 578 (1959). Type:
  Somalia, Serrût Mts near Meid, *Hildebrandt* 1377 (B†, holo.)

CONSERVATION NOTES. This species appears to be widespread and common, and its habitat not
significantly threatened as far as is known. It is rated here as of "least concern" for
conservation.

NOTE. At elevations above 1300 m or so, members of this species become dwarfed in stature,
more densely branched, and have smaller leaves. The syndrome corresponds in general to what
Schumann described as *M. ovata* var. *montana* (although the position of epicalyx bracts does not
seem to be taxonomically significant). Varietal status is not accorded to these plants here
because the suite of characters seems to be environmentally rather than genetically induced.

   M. ovata is easily recognized by its subulate or linear epicalyx bracts that are always
narrower and usually shorter than the sepals. The greyish-green tomentum is also
characteristic. Schumann mentioned longistylous and brevistylous flowers in his species
description implying that there are two different floral morphs in the species. A floral
dimorphism has not been observed in *M. ovata*, although the collection by Faden (*Faden
7/456*) is unusual in its very long style (to 3 mm) and stigmas (to 2 mm). Verdoorn (in
Bothalia 13: 268 (1981)) noted that *M. polygama* I.Verd., known only from Natal, has
"polygamous" flowers; the male flowers with longer petals than the female flowers.

6. **Melhania parviflora** *Chiov.*, Fl. Somala 2: 31 (1932); E.P.A.: 578 (1959);
Blundell, Wild Flow. E. Afr.: t. 289 (1987), excl. syn. *M. taylorii*; Vollesen in Fl. Eth.
2(2): 174, fig. 80.4/2 (1995); Thulin, Fl. Som. 2: 26 (1999). Types: Somalia, near El
Ualud, *Gorini* 485 (FT, syn.), *Gorini* 495 (FT, syn)

Annual or perennial suffrutescent herb or shrub to 0.5 m tall; erect, branches
ascending; young stems canescent to castaneous-tomentose, the hairs stellate with
subappressed arms and often with reddish-brown centres (the coloured base of the
arms); glabrous in age, the stems dark reddish-brown. Leaves narrowly ovate,
narrowly elliptic, or oblong (rarely lanceolate), 1.2–7 cm long, 0.5–3 cm wide, apex
acute, rounded or emarginate, margin very shallowly serrulate to crenulate, base
rounded, velutinous above, paler (often cinereous) below and velutinous, the hairs
above and below stellate, the arms ± ascending and often with reddish-brown centres;
palmately 3–5-nerved from the base; petiole 3–20 mm long, velutinous, the hairs
stellate and often with reddish-brown centres; stipules filiform, 2–6 mm long,
caducous. Inflorescences 2–4-flowered cymes (sometimes appearing subumbellate),
or flowers solitary; peduncle 0.3–2.2 cm long, ± 1 mm wide, velutinous when young,
glabrous in age; pedicels 2–8 mm long, velutinous when young, glabrous in age;
epicalyx bracts lanceolate, narrowly ovate, or ovate (never subulate or linear-
lanceolate), 5–11 mm long, 1.5–4 mm wide, tomentose. Floral buds ovoid, the sepal
tips very briefly free. Sepals narrowly ovate, 4–10 mm long, 1.5–3 mm wide,
sometimes caudate, tomentose without, glabrous within. Petals broadly obovate,
slightly asymmetrical, 5–8 mm long, 4–10 mm wide, ± truncate to slightly rounded
apically, bright to pale yellow, glabrous. Staminal tube 0.5–2 mm long, free portion
of filaments 1–2 mm long, anthers 1–3 mm long, free portion of staminodes 3.5–6 mm
long, glabrous. Ovary ovoid, 2–3 mm long, 2–3 mm in diameter, velutinous; 4 ovules
per locule; style (1–)2–3 mm long; stigma lobes 1–1.5 mm long, erect. Capsules
subglobose to ovoid, 4–7 mm long, 4–8 mm in diameter, tomentose, 2–3(–4) seeds
per locule; seed trigonal to ovoid, 1.5–2 mm long, 1.5–2 mm wide, testa densely
muricate, wihout an elaiosome.

UGANDA. Karamoja District: 1 km from Karita, 10 Oct. 1964, *Leippert* 5097!
KENYA. Northern Frontier District: 40 km on the El Wak-Wajir road, 29 Apr. 1978, *Gilbert &
  Thulin* 1193!; Kilifi District: Kinyoli, 32 km from Malindi, June 1959, *Rawlins* 800!; Kwale
  District: Twiga, 18 Jan. 1964, *Verdcourt* 3950!

TANZANIA. Rufiji District: Mafia Island, Juani Mji, 23 Sep. 1937, *Greenway* 5298!; Kilwa District: Selous Game Reserve, 23 km SW of Kingupira thicket, 23 May 1975, *Vollesen* MRC 2361!
DISTR. **U** 1; **K** 1, 7; **T** 6, 8; Ethiopia and Somalia
HAB. *Acacia - Commiphora* woodland and coastal bushland, on limestone and soils derived from coral; 0–850(–1400) m

CONSERVATION NOTES. This species appears to be fairly widespread and common (15 specimens in the FTEA area alone), and its habitat not significantly threatened as far as is known. It is rated here as of "least concern" for conservation.
NOTE. Thulin (1999: 26), and others, have observed that *M. parviflora* is very close to, and possibly conspecific with, *M. cannabina* Mast., which was described from India. The latter name would have priority if the two names were combined as synonyms.
  Material from **U** 1 (*Lieppert* 5097) and **K** 1 (*Gilbert & Thulin* 1193; *Gillett* 13235), in particular, has longer and narrower (lanceolate) leaves. In this respect the material resembles greatly that of *M. ovata* from Ethiopia and a few specimens from Somalia. The more typical *M. parviflora* is frequently collected in **K** 7 (and Somalia) and has ovate leaves with relatively longer petiole. The lanceolate, narrowly ovate, or ovate (as opposed to subulate or linear) epicalyx bracts readily separate *M. parviflora* from *M. ovata*. The taxon is reported in the Mkomazi checklist (**T** 3), but I have not been able to verify the record.
  The leaves on several specimens (e.g., *Napier* 6362; *G.R. Williams* 828) from coastal Kenya have been eaten by leaf miners or other insects leaving distinctive elongate, brown holes in the epidermis.

7. **Melhania rotundata** *Mast.* in F.T.A. 1: 230 (1868); Baker f. in J.B. 36: 5 (1898); K. Schum. in E.M. 5: 4 (1900); E.P.A.: 579 (1959); Vollesen in Fl. Eth. 2(2): 174, fig. 80.4/3 (1995); Thulin, Fl. Som. 2: 26 (1999). Type: Ethiopia, near Gurrsarfa [Garrsafa], *Schimper* 2286 (K!, holo.; BM!, P, iso.)

Suffrutescent herb or subshrub, 0.3–0.7(–1) m tall; erect, with spreading and ascending branches; young stems cinereous, the hairs stellate with arms ± appressed and sometimes united at a yellow-brown base, glabrescent in age, older stems often reddish. Leaves broadly ovate to round, (1.9–)2.5–6.2 cm long, 1.6–5.7 cm wide, apex rounded to retuse to shortly acute, margin serrate to crenate, base shallowly cordate to truncate, discolorous, pale green to greyish above, paler below, both surfaces tomentose with whitish, stellate hairs with arms ± appressed and often united at a yellow-brown base; palmately 3–7(–9)-nerved from the base; petiole 1–4 cm long, cinereous; stipules filiform, 3–4(–7) mm long, caducous. Inflorescences 2–3-flowered axillary cymes or flowers solitary; peduncle 1.5–5.5 cm long, ± 0.5 mm wide, cinereous; pedicels 5–15 mm long, tomentose; epicalyx bracts lanceolate to narrowly ovate, 7–16(–20) mm long, 2.5–5(–7) mm wide, long acuminate, fused basally, reflexed early, tomentose without and within. Floral buds ovoid, the darkened sepal apices free. Sepals narrowly ovate, 6–15(–18) mm long, 2.5–5 mm wide, caudate, tomentose without, ± glabrous within. Petals broadly obovate, asymmetrical, 8–19 mm long, 6–11.5 mm wide, bright (deep or golden) yellow, glabrous. Staminal tube 1.5–2 mm long, free portion of filaments 1.5–3 mm long, anthers (2–)3–4 mm long, free portion of staminodes 7.5–9 mm long, glabrous. Ovary globose-depressed, ± 4 mm long, 4 mm in diameter, tomentose; 4–6 ovules per locule; style 5–6 mm long, glabrous; stigma lobes 2–3 mm long, recurved. Capsules ellipsoid to almost cylindrical, apically narrowed and ± flattened, 7–14 mm long, 7–12 mm in diameter, tomentose; 4–6 seeds per locule; seed ovoid, 1.5–2 mm long, 1.5–2 mm wide, testa sparsely tuberculate to smooth, without an elaiosome.

KENYA. Northern Frontier District: Wajir, Jan. 1955, *Hemming* 436!; Garissa District: Garissa–Mado Gashi, 26 km from Garissa, 14 Dec. 1977, *Stannard & Gilbert* 1071!; Tana River District: W of road, 19 Dec. 1964, *Gillett* 16437!
TANZANIA. Pare District: Mkomazi Game Reserve, Kamakota Hill, 11 June 1996, *Abdallah et al.* 96/165!
DISTR. **K** 1, 4, 7; **T** 3; Ethiopia and Somalia
HAB. Open to dense *Acacia-Commiphora* woodland or bushland on red sandy soil or laterite; 100–1150 m

Syn. *M. cyclophylla* Mast. in F.T.A. 1: 230 (1868); Baker f. in J.B. 36: 5 (1898); K. Schum. in E.M. 5: 5 (1900); E.P.A.: 576 (1959). Type: Ethiopia, near Gurrsarfa [Garrsarfa], *Schimper* 2205 (K!, holo.; BM!, P, S, iso.)
  M. taylorii Baker f. in J.B. 39: 123 (1901), as *taylori*. Type: Kenya, Mombasa, Freretown, 11 Dec. 1885, *Taylor* s.n. (BM!, holo.)

Conservation notes. This species appears to be fairly widespread and common, and its habitat not significantly threatened as far as is known. It is known from 10 sites in the FTEA area. It is rated here as of "least concern" for conservation.

8. **Melhania randii** *Baker f.* in J.B. 37: 425 (1899); K. Schum. in E.M. 5: 6 (1900); Wild in F.Z. 1(1): 530, t. 99/F (1960); I. Verd. in Bothalia 13: 266–267 (1981). Type: Zimbabwe, Harare [Salisbury], *Rand* 439 (BM, holo.)

Suffrutescent subshrub, to 0.6 m tall; erect, with several slender woody stems arising from a thick, woody rootstock or caudex; young branches finely stellate-tomentose with minute reddish-brown stellate hairs intermingled; glabrescent in age. Leaves narrowly oblong-elliptic or lanceolate-elliptic to ovate-oblong, 2–7 cm long, 0.7–1.5 cm wide, apex subacute to rounded, mucronate or submucronate, margin entire, rarely shallowly dentate in part, base narrowly cuneate to rounded, finely stellate-tomentose above (in some forms early glabrescent), finely and minutely stellate-tomentose with scattered reddish-brown stellate hairs below; palmately 3-nerved from the base; petiole 0.2–1 cm long, ± 0.5 mm wide, tomentose; stipules narrowly lanceolate or subulate, 4–8 mm long, caducous. Inflorescences 2–3-flowered cymes or flowers solitary; peduncle 7 mm long and indistinguishable from the pedicels or up to 4 cm long; pedicels to 1.2 cm long; peduncle and pedicels tomentose; epicalyx bracts narrowly lanceolate (narrowly ovate to ovate-acuminate), 6–9(–13) mm long, 1–3(–6) mm wide (larger bracts found mostly on specimens outside FTEA area), slightly shorter than the sepals, tomentose within and without, and with minute reddish-brown stellate hairs intermixed outside. Floral buds ovoid, the sepal tips confluent. Sepals lanceolate-acuminate, 8–11 mm long, 3–4 mm wide, tomentose and with conspicuous reddish brown stellate hairs without, glabrous within. Petals obovate, asymmetrical, 8–9 mm long, 4.5–7 mm wide, apex oblique, yellow or bright golden yellow, glabrous. Staminal tube ± 1.5 mm long, free portion of filaments ± 1 mm long, anthers 2–3 mm long, free portion of staminodes 3–5 mm long, glabrous. Ovary globose to ovoid, 2–2.5 mm long, 2–3 mm in diameter, velutinous, with stellate hairs and straight bristly hairs especially in upper portion; 6 ovules per locule; style ± 1 mm long, glabrous; stigma lobes ± 1.5 mm long, erect to spreading. Capsules narrowly ovoid, 7–10 mm long, 8–10 mm in diameter, stellate-pubescent with some hairs scale-like, 1–3(–6) seeds per locule (1 seed per locule in FTEA specimen); seed ± ovoid, 2–2.5 mm long, 1–1.5 mm wide, testa smooth, without an elaisome.

Tanzania. Njombe District: N Upangwa, N side of Shawimbi Mt, Kijombo, 20 Aug. 1982, *Leedal* 7134!
Distr. **T** 7; Malawi, Mozambique, Zimbabwe, and northern South Africa
Hab. Not stated by the collector, probably open woodland; 2100 m

Syn. [*Melhania prostrata* sensu Eyles in Trans. Roy. Soc. S. Afr. 5: 417 (1916); Hopkins et al., Comm. Veld Fl.: 73, fig. (1940), *non* DC.]
  [*M.* sp.; Eyles in Trans. Roy. Soc. S. Afr. 5: 418 (1916). Material cited: "Mazoe, *Eyles* 411"]

Conservation notes. This species though known from a single specimen in the FTEA area, is common further south (23 specimens in the Flora Zambesiaca area), is widespread and common, and its habitat not significantly threatened as far as is known. It is rated here as of "least concern" for conservation.
Notes. The stout caudex and collectors' notes indicate that *M. randii* is adapted to growing in vegetation that is periodically burned.

Our material of *M. randii* differs from that described by Verdoorn (1981) in having fruits with 1-seeded locules, in which as many as 3 additional aborted ovules can be detected. Verdoorn did note that this species has a discontinuous distribution, and the isolation of populations undoubtedly contributes to their diverging one from the other with respect to various characters.

M. randii is very similar to *M. prostrata* DC., which is more widespread in southern Africa, but the former may be distinguished from the latter by leaf vesture (finely and minutely stellate-pubescent above versus sparsely pubescent with mostly long, simple subappressed hairs), epicalyx bract length (6–9(–13) versus 8–15 mm) and shape (lanceolate-acuminate versus narrowly ovate-acuminate), and seed testa (smooth versus reticulate). *M. randii* differs from *M. integra* I.Verd., the only other entire-margined *Melhania* in Africa, by its shorter style (± 1 versus 6–8 mm long) and shorter petals (8–9 versus ± 16 mm long).

9. **Melhania volleseniana** *Dorr* **sp. nov.** a *M. ovata* Cav. indumento lanato densiore, stipulis longioribus persistentibusque, segmentis epicalycis longioribus lateralibusque et sepalis longioribus differt. Typus: Kenya, Northern Frontier District: Kailongol Mts, *Mathew* 6818 (K!, holo.)

Annual or perennial herb to 0.5 m tall; erect; stems whitish lanate. Leaves ovate to widely elliptic, 2.5–7 cm long, 1.5–4.5 cm wide, apex subacute, apiculate, or truncate, margin dentate to serrate, base truncate, weakly to strongly discolorous, green to greyish pubescent above, densely greyish-white lanate below, palmately 3–5-nerved from the base; petiole 0.5–3 cm long, whitish lanate; stipules filiform, 0.8–2 cm long, persistent, sparsely to densely whitish-stellate pubescent. Inflorescences 2–3-flowered cymes, or flowers solitary; peduncle 1–2(–3.5) cm long, ± 1.5 mm wide; pedicels to 1 cm long; peduncle and pedicels tomentose to lanate; epicalyx bracts subulate to linear or narrowly lanceolate, (8–)10–20 mm long, 1–2(–3) mm wide, most ± as long as the the sepals or slightly longer, tomentose, erect. Floral buds narrowly ovoid, the sepal apices free. Sepals subulate, (9–)11–18 mm long, 3–4 mm wide, long-acuminate, tomentose to lanate without and within above, glabrous toward the base, individual hairs long-stalked stellate. Petals obovate, slightly asymmetrical, 6–10 mm long, 3.5–4 mm wide, yellow or pale yellow, glabrous. Staminal tube 1–1.5 mm long, free portion of filaments ± 1.5 mm long, anthers ± 1.5 mm long, free portion of staminodes 4–6 mm long, glabrous. Ovary subglobose, 3.5–4 mm long, 3.5–4 mm in diameter, densely white-velutinous; 6 ovules per locule; style 2–3 mm long, sometimes with a few scattered white stellate hairs; stigma lobes 1–1.5 mm long, erect, becoming recurved. Capsules ovoid, 7–9 mm long, 6–9 mm in diameter, tomentose; 4(–6) seeds per locule; seed trigonal with two flat and one rounded surface, 1.5–2 mm long, 1–1.5 mm wide, tuberculate, without an elaiosome. Fig. 14/10–12, p. 85.

KENYA. Northern Frontier District: Kailongol Mts, 15 June 1971, *Mathew* 6818!; Turkana District: Lokitunyalla, July 1932, *Champion* T.146!; Meru District: Meru National Park, Kiolu R. crossing, 25 Dec. 1972, *Gillett* 20145!
DISTR. **K** 1, 2, 4; Somalia and southern Ethiopia
HAB. *Acacia-Commiphora* bushland and *Combretum* wooded grassland on lava or stony soil; 290–700 m

SYN. ["*Melhania* sp. = *Thesiger* 1945" in Vollesen in Fl. Eth. 2(2): 173 (1995); Thulin, Fl. Som. 2: 24 (1999)]

CONSERVATION NOTES. This species appears to be on the brink of the rating "threatened" since it is known from only ± 15 sites: only a single collection is known from Somalia and nine from the FTEA area. Its habitat is threatened in at least part of its range in Kenya (**K** 1: see notes under *Hermannia pseudathiensis*). Accordingly it is rated here as "near threatened".
NOTE. Related to *M. ovata*, but differing in having a denser, lanate indumentum, longer and persistent stipules, longer and wider epicalyx bracts, and longer sepals. Vollesen (1995) also considered the hairy style to be a distinguishing character, but Thulin (1999) correctly maintained that this was an inconstant character.

Some of the Ethiopian material (e.g., *Gilbert et al.* 9088) has relatively smaller, more ovate-lanceolate leaves, but otherwise agrees well with the rest of the collections examined.

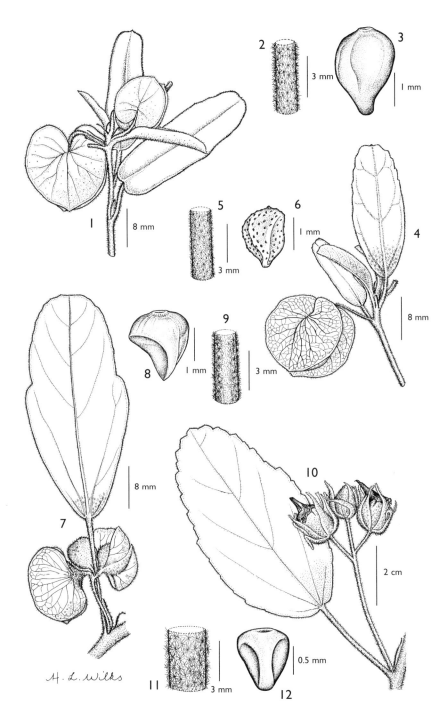

FIG. 14. *MELHANIA PRAEMORSA* — **1**, habit; **2**, stem indumentum; **3**, seed. *MELHANIA SUBSTRICTA* — **4**, habit; **5**, stem indumentum; **6**, seed. *MELHANIA LATIBRACTEOLATA* — **7**, habit; **8**, stem indumentum; **9**, seed. *MELHANIA VOLLESENIANA* — **10**, habit; **11**, stem indumentum; **12**, seed. 1–3 from *Gillett* 13381; 4–6 from *Gilbert & Thulin* 1345; 7–9 from *Gilbert & Thulin* 1597; 10–12 from *Mathenge* 76. Drawn by Hazel Wilks.

10. **Melhania phillipsiae** *Baker f.* in J.B. 36: 4 (1898); K. Schum. in E.M. 5: 16 (1900); E.P.A.: 578 (1959); Vollesen in Fl. Eth. 2(2): 176, fig. 80.4/7 (1995); Thulin, Fl. Som. 2: 27 (1999). Types: Somalia, Wagga Mt, 1897, *Lort Phillips* s.n. (BM!, syn.; K [as K37]!, isosyn.); Somalia, Soksoda, no date, *Lort Phillips* s.n. (BM!, syn.) - the two syntypes at BM are mounted on the same sheet

Suffrutescent perennial herb or subshrub, to 2 m tall; erect; young stems tomentose to lanate. Leaves ovate to broadly ovate, 2.5–9(–12) cm long, 2–6(–8.5) cm wide, apex rounded to truncate, margin crenate to irregularly serrate, base subcordate; tomentose above and below; palmately 3–5-nerved at the base; petiole 1.5–5.5 cm long, ± 0.2 cm wide, tomentose; stipules linear, 0.3–2 cm long, persistent. Inflorescences 2–6-flowered axillary cymes; peduncle 1.5–4(–5) cm long, ± 0.2 cm wide; pedicels to 1.5 cm long; peduncle and pedicels tomentose to lanate; epicalyx bracts widely cordiform or reniform, 1–2.2 cm long, 1.2–2.5 cm wide (in flower), accrescent, 1.7–4 cm long, 2–5 cm wide (in fruit), apex acute or shortly acuminate, briefly united at base, basal sinus 3–10 mm deep, basal lobes not overlapping, tomentose when young, becoming scarious and reticulate-veined in fruit. Floral buds obscured by epicalyx bracts, widely ovoid, the sepal tips free. Sepals lanceolate, 9–12 mm long, 3–3.5 mm wide (to 25 mm long, 5 mm wide in fruit), acuminate, apex often recurved, densely pilose without, glabrous within except for pilose apex. Petals obovate, asymmetrical, 7–10 mm long, ± 5 mm wide, yellow, glabrous. Staminal tube 1–1.5 mm long, free portion of filaments 1–1.5 mm long, anthers 1.5–2 mm long, free portion of staminodes ± 4.5 mm long, glabrous. Ovary ovoid, 3.5–4 mm long, 3.5–4 mm in diameter, tomentose, 3 ovules per locule; style 1–3 mm long, glabrous; stigma lobes ± 2 mm long, recurved. Capsules ovoid to subglobose, 7–10 mm long, 7–10 mm in diameter, densely stellate-pubescent, papillate from the persistent bases of the stellate hairs, 2–3 seeds per locule; seed angular, 3–4 mm long, ± 2.5 mm wide, testa tuberculate, without an elaisome.

KENYA. Northern Frontier District: 53 km SW of Mandera on El Wak road, 30 May 1952, *Gillett* 13401!
DISTR. **K** 1; Niger, Chad, Eritrea, Djibouti, Ethiopia and Somalia; also in southern Egypt (Jebel Elba), Arabia
HAB. *Acacia-Commiphora* scrub; ± 390 m

SYN. *M. denhamii* R.Br. var. *grandibracteata* K.Schum. in Ann. Istit. Bot. Roma 7: 34 (1897); Baker in J.B. 36: 4 (1898); E.P.A.: 577 (1959). Types: Somalia [Ethiopia?], near Menehan, Balambel, *Riva* 440 (B†, syn.; FT, iso.), *Riva* 441 (B† syn.; FT, iso.)
  M. grandibracteata (K.Schum.) K.Schum. in E.M. 5: 15–16, t. 1/B (1900); Kuchar in CRDP Tech. Rept. Ser. 16: 95 (1986)
    M. fiorii Chiov. in Ann. Bot. (Roma) 10: 383 (1912); E.P.A.: 577 (1959). Types: Eritrea, Habab to Grammé, *Pappi* 7955 (FT, syn.); Habab to Kub-Kub, *Pappi* 8022 (FT, syn.); Habab to Magbar, *Pappi* 8199 (FT, syn.)

CONSERVATION NOTES. This species appears to be very widespread and fairly common, with 15 specimen-sites in Ethiopia-Somalia alone, although only a single site is known in the FTEA area. Its habitat is not generally significantly threatened as far as is known, although it is at the Kenyan site (see under *Hermannia pseudathiensis*). It is rated here as of "least concern" for conservation.
NOTES. *M. hiranensis* Thulin, known only from Somalia, is closely related to *M. phillipsiae*. The latter differs from the former in having 2–6-flowered (versus solitary flowers or 2-flowered) inflorescences, pilose (versus subglabrous) sepals, and 2–3-seeded (versus 1-seeded) locules. Additionally, the leaves, epicalyx segments, sepals, petals, capsules and seed are all larger in *M. phillipsiae* than in *M. hiranensis*.
    The syntypes of *M. denhamii* var. *grandibracteata* are as cited by Schumann (1897). However, Cufodontis in E.P.A. (1959: 577) wrote *Robecchi-Bricchetti* 1140 ("typus"), and Vollesen (1995) and Thulin (1999) cited *Robecchi-Brichetti* 440 & 441 as syntypes.

11. **Melhania latibracteolata** *Dorr* **sp. nov.** *M. kelleri* Schinz affinis sed segmentis epicalycis saltem 5 mm latioribus quam longioribus et sepalis pilosis omnino notabilis. Typus: Ethiopia, Sidamo, Melca Guba, at the Dawa Parma River, *Friis, Mesfin & Vollesen* 2971 (K!, holo.)

Suffrutescent herb or small shrub to 0.4 m tall; erect, branches dense, ascending; young stems lanate, the hairs whitish, long-armed stellate or tufted-stellate, glabrous in age. Leaves ovate to elliptic, 1.5–7.5 cm long, 0.9–3.2 cm wide, apex rounded to emarginate, margin shallowly and distantly dentate to crenate, base rounded to cuneate, larger leaves occasionally with 2–4 shallow lateral lobes, upper surface very densely cinereous to whitish-tomentose when young, becoming sparsely tomentose, paler greyish-green tomentose below, the hairs stellate; palmately 3–5-nerved from the base; petiole 0.5–3.2 cm long, apex ± pulvinate, whitish tomentose, the hairs tufted-stellate; stipules filiform to subulate, 4–8 mm long, reddish-brown, sparsely pubescent. Inflorescences 2–5-flowered cymes (appearing subumbellate); peduncle 0.5–1.5 cm long, ± 0.5 mm wide; pedicels 2–5 mm long, ± filiform; peduncle and pedicels tomentose with tufted-stellate hairs; epicalyx bracts very broadly ovate to widely reniform, 5–7 mm long, 8–12 mm wide (in flower), tomentose, accrescent, 10–11(–15) mm long, 15–20(–30) mm wide (in fruit), apex rounded to very shortly acute, basal sinus shallow, the basal lobes rounded, never overlapping, sparsely tomentose, especially toward the base, membranous. Floral buds obscured by the epicalyx bracts. Sepals narrowly ovate to lanceolate, 5–7 mm long, 1–2 mm wide, accrescent, becoming membranous in fruit, pilose without, glabrous within. Petals obovate, asymmetrical, 4–5 mm long, 3.5–4 mm wide, (pale) yellow. Staminal tube ± 1 mm long, free portion of filaments ± 1.5 long, anthers ± 0.5 mm long, free portion of staminodes 2–2.5 mm long, glabrous. Ovary globose to ovoid, 3.5–4 mm long, 3–3.5 mm in diameter, velutinous; 3 ovules per locule; style 0.5–1 mm long, glabrous; stigma lobes ± 1.5 mm long, erect to ascending. Capsules depressed globose to ovoid, 3–4.5 mm long, 4–5 mm in diameter, stellate-pubescent to velutinous; 1–2 seeds per locule; seed ovoid, ± trigonal, flattened on one side, with an umbonate projection, 2–2.5 mm long, 1.5–2 mm wide, testa smooth, without an elaisome. Fig. 14/7–9, p. 85.

KENYA. Northern Frontier District: 43 km on the Wajir–El Wak road, 29 Apr. 1978, *Gilbert & Thulin* 1166! & 48 km on the Ramu–El Wak road, May 1978, *Gilbert & Thulin* 1597!; Della [Dela], 23 June 1951, *Kirrika* 70!
DISTR. **K** 1; Ethiopia and Somalia
HAB. *Acacia-Commiphora* bushland, often on limestone soils; 450–900 m

SYN. ["*M.* sp. = *Friis et al.* 2971" in Vollesen in Fl. Eth. 2(2): 176 (1995)]
    [*M. denhamii* sensu Thulin, Fl. Som. 2: 27 (1999) *pro parte, non* R.Br.]

CONSERVATION NOTES. This species is here rated as VU B2a, b(iii), that is vulnerable to extinction being known from 10 sites or less (ten are known), with an area of occupancy of less than 20,000 km² and with reduction of habitat quality in at least part of its range (see notes regarding **K** 1 under *Hermannia pseudathiensis*).
NOTE. Similar to *M. kelleri* Schinz except that the epicalyx bracts are at least 5 mm wider than long and the sepals are pilose all over.
    The flowers are said to open in the evening (*Gilbert & Thulin* 1597).

12. **Melhania substricta** *Dorr* **sp. nov.** *M. muricata* Balf.f. affinis sed indumento minus adpresso, stipulis pedicellisque longioribus, segmentis epicalycis parvioribus cum sinibus laeve profundioribus, et sepalis brevioribus notabilis. Typus: Ethiopia, Ogaden, Kebre Dahar, *J. de Wilde* 5986 (K!, holo., WAG, iso.)

Suffrutescent subshrub, to 50 cm tall; erect, the branches ascending; young stems pale brown tomentose, the hairs mixed, sessile-stellate or tufted stellate, or somewhat brownish-centered peltate-stellate hairs, glabrous in age. Leaves narrowly elliptic (FTEA area) to ovate, 0.9–3.6 cm long, 0.8–2.6 cm wide, apex rounded to emarginate, sometimes mucronate, margin shallowly serrulate to crenulate especially toward the apex, base rounded; somewhat discolorous, greenish tomentose above, becoming sparsely so, the hairs tufted stellate, paler below and more densely whitish tomentose (less so with age); palmately 3–5-nerved from the base; petiole 0.5–1 cm long, pale brown tomentose; stipules filiform, 1–5 mm long, reddish brown, sparsely pubescent, persistent. Flowers solitary, rarely 2–flowered cymes; pedicel 8–20 mm long, pubescent with stellate hairs; epicalyx bracts cordate, 8–13 mm long, 8–12 mm wide (in flower), tomentose to pubescent with whitish stellate hairs, accrescent, 16–26 mm long, 13–25 mm wide (in fruit), apex rounded or acute, base cordate, sinus $\frac{1}{4}$ to $\frac{1}{3}$ the height of the bract (in fruit), basal lobes of sinus sometimes overlapping, becoming ± chartaceous to membranous, sparsely pubescent in fruit. Floral buds concealed by the epicalyx bracts. Sepals lanceolate, 3.5–4.5 mm long, ± 1.5 mm wide, acute, pubescent without, glabrous within, chartaceous. Petals obovate, asymmetrical, 4.5–5 mm long, 5–6 mm wide, yellow, glabrous. Staminal tube 0.5–1 mm long, free portion of filaments 1–1.5 mm long, anthers 1–1.5 mm long, free portion of staminodes 2.5–3 mm long, glabrous. Ovary ovoid to pyriform, 1.5–2 mm long, 1.5–2 mm in diameter, velutinous; 3 ovules per locule; style 0.5–1 mm long, glabrous; stigma lobes ± 1.5 mm long, reflexed. Capsules spheroid to widely or narrowly ovoid, 4–8 mm long, 5–6 mm in diameter, ± 5-lobed, velutinous, 3 seeds per locule; seed trigonal, at least one side flattened, 1.5–2 mm long, ± 1.5–2 mm wide, testa tuberculate, without an elaiosome. Fig. 14/4–6, p. 85

KENYA. Northern Frontier District: 5–10 km SSE of Ramu, 2 May 1978, *Gilbert & Thulin* 1345!
DISTR. **K** 1; Eritrea, Ethiopia, and Somalia; also in Yemen
HAB. *Acacia-Commiphora* woodland and bushland, often on limestone slopes; 400–500 m

SYN. ["*M.* sp. = *Gilbert et al.* 7589" in Vollesen in Fl. Eth. 2(2): 175 (1995)]
    [*M. muricata* sensu Thulin, Fl. Som. 2: 29 (1999), *pro parte*]

CONSERVATION NOTES. This species is here rated as VU B2a, b(iii), that is vulnerable to extinction being known from 10 sites or less (eight are known), with an area of occupancy of less than 20,000 km^2 and with reduction of habitat quality in at least part of its range (see notes regarding **K** 1 under *Hermannia pseudathiensis*).
NOTE. *M. substricta* is close to, but clearly distinct from, *M. muricata* Balf.f. from Arabia, Socotra, and Somalia. The latter species has a more appressed indumentum, shorter stipules (to 2 mm) and pedicels (to 7 mm), bracteoles up to 20 × 30 mm with a relatively shallower sinus (less than 5 mm deep), longer sepals, and 2–3-seeded locules.
    M. substricta also is similar to *M. stipulosa* J.R.I.Wood, which is known from the Horn of Africa and Arabia. The two differ with respect to pubescence (± sessile stellate hairs with ascending arms versus stellate hairs with appressed arms, dark center and short stalks) (in this respect *M. stipulosa* is closer to *M. muricata*); in the larger sinus of the epicalyx bracts with overlapping basal lobes (versus non-overlapping basal lobes); leaf shape (narrowly elliptic to ovate versus narrowly oblong); and short caducous (versus long persistent) stipules.

13. **Melhania praemorsa** Dorr **sp. nov.** *M. denhamii* R.Br. affinis sed foliis 3plo longioribus quam latioribus et inferioribus segmentis epicalycis aliquando connatis notabilis. Typus: Ethiopia, Ogaden, 26 km E of Wadere, *Hemming* 1465 (K!, holo.)

Perennial, suffrutescent herb or subshrub to 60 cm tall; erect; young stems whitish tomentose to canescent or lanate, glabrescent in age. Leaves lanceolate to narrowly ovate or narrowly oblong, 1.5–6.7 cm long, 0.5–2.2 cm wide, apex subacute to retuse, sometimes mucronulate to mucronate, margin entire or remotely crenate, or denticulate (rarely coarsely dentate) in upper half, base truncate to subcordate, velutinous above, densely softly whitish pubescent below with scattered dark-based stellate hairs; palmately 3(–5)-nerved from the base; petiole 0.4–1.3 cm long, whitish tomentose; stipules subulate to filiform, 0.3–1 cm long, caducous, sometimes persisting. Inflorescences 2–3(–4)-flowered cymes or flowers solitary; peduncle 0.5–4 cm long, ± 0.5 mm wide; pedicels 0.5–1 cm (–3.5 cm in solitary flowers) long; peduncle and pedicels pilose to tomentose; epicalyx bracts widely cordiform to reniform, 0.5–1.1 cm long, 0.6–1.3 cm wide (in flower), accrescent, 1.2–2.5 × 1–2.5 cm (in fruit), acute, free or united at the base, basal sinus 3–8 mm deep, tomentose, becoming pilose in fruit, with short-stalked stellate hairs, chartaceous. Floral buds obscured by the epicalyx bracts, ovoid to conical, sepal apices confluent. Sepals ovate to lanceolate, 3–7(–9 in fruit) mm long, 1–3 mm wide, pilose to pubescent. Petals obovate, strongly asymmetrical, 3.5–5 mm long, 3–4 mm wide, yellow, glabrous. Staminal tube ± 0.5 mm long, free portion of filaments ± 1 mm long, anthers ± 0.5–1 mm long, free portion of staminodes 2–4 mm long, glabrous. Ovary obpyriform, ± 2 mm long, ± 2 mm in diameter, stellate-pubescent; 1(–2) ovules per locule; style 1–2 mm long, glabrous; stigma lobes ± 1 mm long, recurved. Capsules obovoid to obpyriform (narrowest at the base), 4–5 mm long, 3.5–4 mm in diameter, stellate pubescent; 1(–2) seeds per locule; seed trigonal, 2–3 mm long, 1.5–2 mm wide, testa smooth, with a pronounced umbonate projection. Fig. 14/1–3, p. 85.

KENYA. Northern Frontier District: 13–22 km SSW of El Wak on Wajir road, 29 May 1952, *Gillett* 13381!
DISTR. **K** 1; Somalia and Ethiopia
HAB. *Acacia-Commiphora* open bushland on red, sandy soil; ± 400 m

SYN. ["*M.* sp. = *Hemming* 1465" in Vollesen in Fl. Eth. 2(2): 176 (1995)]
 [*M. denhamii* sensu Thulin, Fl. Som. 2: 27–28 (1999) pro parte, *non* R.Br.]

CONSERVATION NOTES. This species is here rated as VU B2a, b(iii), that is vulnerable to extinction being known from 10 sites or less (ten are known), with an area of occupancy of less than 20,000 km² and with reduction of habitat quality in at least part of its range (see notes regarding K1 under *Hermannia pseudathiensis*).
NOTES. Clearly allied to *M. denhamii* R.Br., *M. praemorsa* can be distinguished by its leaves that are 3 or more times long as wide and by its epicalyx bracts that are sometimes united at the base.

DOUBTFUL SPECIES

14. **Melhania polyneura** *K.Schum.* in E.M. 5: 8 (1900); T.T.C.L.: 601 (1949). Type: Tanzania, Mwanza District: Kagehi [Kayenzi], *Fischer* 64 (B†, holo.)

Herb, to 20 cm tall; becoming woody and probably suffruticose; leaves ovate-oblong, 1–4 cm long, 1–3 cm wide, serrulate, tomentose; palmately 5-nerved from the base and with ± 8 pairs of lateral nerves; petiole 1–2 cm long. Inflorescences 4-flowered (cymes?), not or scarcely longer than petiole; epicalyx bracts ovate or ovate-oblong, 0.7 cm long; sepals 0.9 cm long; petals 1 cm long, yellow.

TANZANIA. Mwanza District: Kagehi, *Fischer* 64; Tabora (fide T.T.C.L.)
DISTR. **T** 1; endemic?
HAB. No data

NOTE. This name is of uncertain application. Schumann (1900) placed it in *M.* subgen. *Broteroa* (Cav.) K.Schum., and near *M. ovata* and *M. denhardtii* in his key. He distinguished *M. polyneura* from these two species by the number of pairs of lateral nerves above the base of the leaf blade; 8–9 pairs in *M. polyneura* versus 4–5 pairs in the other two species.
 A sketch of an epicalyx bract and a leaf at the BM purports to be based on the type of *M. polyneura*. The epicalyx bract also is described as "bracts oblong pointed." The leaf blade is rather nondescript, ovate to ovate-lanceolate in shape (the nerves are not shown), and it is described as "cinereous tomentose." Among the FTEA *Melhania*, *M. polyneura* would seem to most closely resemble *M. parviflora* Chiov., but in the absence of authentic material I am reluctant to formally consider the names synonyms.
 It is unclear why Brenan (T.T.C.L.) cited Tabora in the distribution of this species. There is no indication that he examined the type.

10. **HARMSIA**

K.Schum. in Ann. Ist. Bot. Roma 7(1): 35 (1897) &. in E. & P.Pf., Nachtrage 1: 240 (1897) &. in E.M. 5: 17 (1900); Jenny et al. in Taxon 48: 3–6 (1999)

Aethiocarpa Vollesen in K.B. 41: 959 (1986)

Shrubs or shrublets. Leaves simple, entire or crenate, palmately nerved at the base, pinnately nerved above; petiolate; stipules persistent. Inflorescences axillary, few-flowered, umbelliform or racemoid cymes, or flowers solitary; peduncle and pedicels expanding in fruit; epicalyx bracts 3, entire. Floral buds ovoid to linear-ovoid. Flowers bisexual. Sepals 5, briefly joined at the base, sometimes persistent. Petals 5, symmetric to asymmetric, yellow to orange, glabrous, persistent and papery in fruit; stamens and staminodes in a single whorl at the apex of a short staminal tube; stamens antipetalous, 15 in fascicles of 3 (1 long and 2 short), each fascicle alternating with a ligulate or linear staminode; anthers di-thecate, extrorse; pollen echinate. Ovary syncarpous, unilocular or seemingly 2-locular by virtue of a false septum connected on one side and merely connate on the other, stellate-pubescent; 2 ovules per locule, or 1 by abortion; ovules basal, ascending; style simple, clavate, stigma undifferentiated or briefly 4-lobed. Capsules 5-angled, with 2 rows of bristles or prickles at the angles, pubescent, endocarp glabrous, indehiscent (?) or tardily dehiscent, opening septicidally from the base, often dispersed as a whole; 1–2 seeded. Seed without endosperm; cotyledons bifid, contorted.

Two species found in North East Africa (Somalia, Ethiopia and Kenya).

Harmsia sidoides *K.Schum.* in Ann. Ist. Bot. Roma 7(1): 35 (1897) & in E. & P.Pf., Nachtrage 1: 240 (1897) & in E.M. 5: 17–18, t. 1, fig. G/a-i (1900); Cufodontis in Miss. Biol. Borana, Racc. Bot. 4: 138 (1939); E.P.A.: 579 (1959); K.T.S.: 548 (1961); K.T.S.L.: 165, fig., map (1994); Vollesen in Fl. Eth. 2(2): 170, fig. 80.3:1–8 (1995); Thulin, Fl. Som. 2: 23, fig. 10 (1999). Types: Ethiopia, near Giacorse well, *Riva* 451 (FT, syn.); near Ualeme, *Riva* 1538 (FT, syn.); Web valley, *Riva* 447 (FT, syn.)

Shrub, to 2 m tall, erect with virgate branchlets; young stems densely yellowish to greyish stellate-pubescent, glabrescent. Leaves elliptic to obovate or subovate, (1–)2–5(–6.5) cm long, 0.5–2.8(–3.5) cm wide, apex shortly acute or rounded, often emarginate or retuse and mucronate, margin dentate-serrate toward the apex, entire or subentire toward the base, base rounded to slightly cordate or subcordate; densely stellate puberulous below, less so above; palmately 3–5-nerved from the base, pinnately nerved above; petiole 2–6(–10) mm long, densely stellate puberulous; stipules linear-subulate, often longer than the petiole, 5–10(–17) mm long, persistent. Inflorescences axillary, 2–4(–5)-flowered umbelliform or racemoid cymes, or flowers solitary; peduncle (5–)8–19(–40) mm long, usually erect; pedicels 5–23 mm long, often pendulous, articulated; peduncle and pedicels stellate-pubescent; epicalyx bracts linear, ± 5 (–8) mm long, borne immediately above the pedicel articulation and removed from the calyx. Floral buds 4–5 mm long, 2–3 mm wide, apiculate, the tips of the sepals ± free, ± 2 mm long. Sepals lanceolate to narrowly ovate, 4–8 mm long, 1–2 mm wide, densely stellate puberulous outside, persistent in fruit. Petals orange or yellow, obovate to obdeltate, asymmetric, 5–7.5(–8) mm long, 4.5–6 mm wide, apex truncate, narrowed to the base, glabrous, persisting on young fruit. Staminal tube 1–1.5 mm long, free portion of long filament ± 0.5–1 mm long, free portion of short filaments 0.3–0.5 mm long, anthers 0.7–1 mm long, staminodes ligulate, free portion 2.5–3 mm long, papillate. Ovary globose to botuliform, ± 1 mm long, ± 0.5 mm in diameter, stellate pubescent; style simple, 1–1.5(–4) mm long, bearing a few appressed simple hairs; stigma clavate, stigmatic surface ± 0.5 mm long. Capsules ellipsoid to obovoid, 4–6 mm long, 3–5 mm in diameter, woody, densely

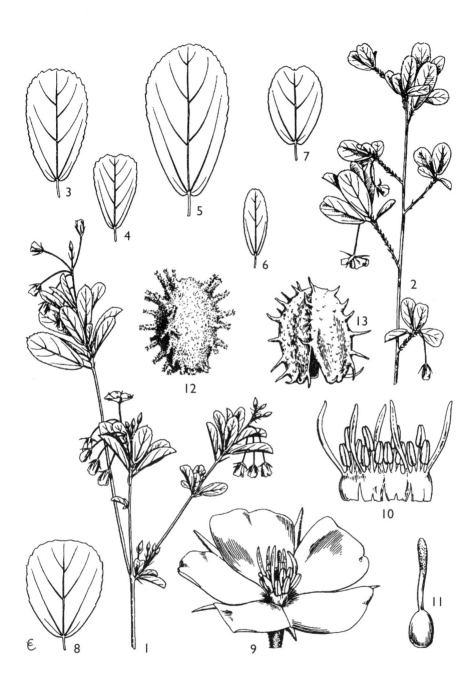

Fig. 15. *HARMSIA SIDOIDES* — **1**, habit, in flower × ²/₃; **2**, habit in fruit ²/₃; **3–8**, leaves, showing variation × 1; **9**, flower × 4; **10**, androecium, opened × 8; **11**, gynoecium × 8; **12**, young fruit × 6; **13**, old, dehisced fruit × 6. 1 from *Friis et al.* 959; 2–4 & 10, 11 from *Thulin et al.* 3586; 5 & 9 from *Friis et al.* 2795; 6 from *Ash* 809; 7 from *Keller* s.n.; 8 &12 from *Ellis* 387; 13 from *Gilbert et al.* 7857. Reproduced from F.E. 2(2): t. 80.3. Drawn by Eleanor Catherine.

stellate-hispid, with 5 longitudinal ridges, each ridge with two rows of short, 1.5–2.3 mm long, stellate-pubescent prickles; seed reddish-brown, obovoid, 3–4 mm long, 2–2.3 mm wide, glabrous, raphe conspicuous. Fig. 15, p. 91.

Kenya. Northern Frontier District: Baldesa, Dandu, 2 Apr. 1952, *Gillett* 12671!; & between Banessa & Ramu, 24–27 km E of Banessa on Ramu road, 22 May 1952, *Gillett* 13261!; & 40 km on the El Wak–Wajir road, 29 Apr. 1978, *Gilbert & Thulin* 1202!

Distr. **K** 1; Somalia, Ethiopia

Hab. *Acacia-Commiphora* woodland and bushland on limestone and on soil overlying limestone; 450–850 m

Syn. *H. emarginata* Schinz in Bull. Herb. Boiss., sér. 2, 2: 1006 (1902); E.P.A.: 579 (1959). Type: Somalia, Abdallah [El Abdalla], 1891, *Keller* s.n. (Z, holo., K!, iso., BM!, sketch of type)
    H. kelleri Schinz in Bull. Herb. Boiss., sér. 2, 2: 1006 (1902), *nom. nud., sphalm.* pro *H. emarginata*
    H. microblastos K.Schum. in E.J. 33: 308 (1903); E.P.A.: 579 (1959); Kuchar in CRDP Tech. Rept. Ser. 16: 94 (1986). Type: Ethiopia, Djaro in Borun, *Ellenbeck* 2059 (B†, holo.)

Conservation notes. This species appears to be fairly widespread and fairly common, and its habitat not significantly threatened as far as is known. Four specimen sites are known from Kenya, five from Somalia and 11 from Ethiopia. It is rated here as of "least concern" for conservation.

11. **NESOGORDONIA**

Baillon, Bull. Mens. Soc. Linn. Paris 1(70): 555 (1886); L.C. Barnett, The Systematics of the genus *Nesogordonia* (Sterculiaceae), unpublished dissertation, University of Texas, Austin (1988)

Cistanthera K.Schum. in E. & P. Pf., Nachtr. zum II-IV: 234 (1898)

Trees, rarely shrubs; bole straight, unbranched for most of its length, narrow. Leaves clustered toward the shoot apex, petiolate, simple, stellate-pubescent, rarely also with glandular and simple hairs, sometimes with domatia in the vein axils below, coriaceous; stipules fugaceous. Inflorescences axillary (rarely terminal) cymes; epicalyx bracts 3, fugaceous. Flowers bisexual, actinimorphic, 5-merous. Calyx 5-lobed, the lobes united briefly at the base, densely stellate-pubescent. Petals contorted, usually fleshy or coriaceous; androecium 2-whorled, contorted, external whorl of 10–25 stamens in 5 antisepalous fascicles, internal whorl of 5 antipetalous, free staminodes (occasionally stamens); stamens laminar; anthers linear, along the margin of the stamen; staminodes laminar, geniculate, usually fleshy. Ovary 5-locular, stellate-pubescent, ovules subapical, 2 per locule, collateral; styles 5, connate and convolute; stigma lobes 5, conical, fleshy, with a recurved stigmatic surface. Capsules 5-valved, loculicidal, woody, obconic to ± spheroid, frequently 5-costulate, frequently flattened at the apex and with a well-defined rim, surface generally stellate-pubescent; seeds 2 per locule, each bearing a membranaceous, unilateral wing toward the base.

A genus of 19 species; 3 in tropical Africa, 1 in the Comoro Islands, and 15 in Madagascar

1. Leaves ovate, 2.5–8.1 cm long, 1.1–3.8 cm wide; stamens 10, in 5 fascicles of 2, each fascicle consisting of stamens in the same whorl; capsule 1.2–2.5 cm long .... 1. *N. holtzii*
   Leaves elliptic to obovate, 7–14.5 cm long, 3.2–6.8 cm wide; stamens 15, in 5 fascicles of 3, each fascicle consisting of 1 external and 2 internal stamens; capsule 3.1–3.5 cm long .... 2. *N. kabingaensis*

1. **Nesogordonia holtzii** (*Engl.*) *L.C.Barnett & Dorr* in K.B. 55: 985 (2000). Type: Tanzania, Uzaramo District: Pugu Forest Reserve, *Semsei* 1751 (EA!, neo.; BR!, EA!, K!, isoneo., chosen by Verdcourt 1956)

Tree to 30 m tall, rarely a large shrub; bole straight to 25 m, slightly buttressed or fluted at the base; outer bark white to greyish to brown, and smooth, wrinkled, rough, or slightly flaking; crown spreading young branchlets grey, brown or tan, striate to wrinkled with prominent projecting leaf scars, pubescent with simple or fasciculate and flattened-stellate hairs; becoming glabrous in age. Leaves ovate, 2.5–8.1 cm long, 1.1–3.8 cm wide, apex acuminate, usually mucronate, margin entire to slightly crenate toward the apex, and very slightly revolute, base rounded; sparsely stellate-pubescent below, glabrescent; glossy and glabrous above; below with conspicuous domatia in the vein axils; petiole 0.4–4 cm long, pulvinate, hispid; stipules fugacious. Inflorescences 1–2-flowered axilary cymes; peduncle and pedicels stellate-pubescent; peduncle 0.3–3.8 cm long; pedicels 1–10 cm long at anthesis, often articulated 1–3 mm below the flower. Floral buds ovoid, 6–8 mm long, 3–4 mm wide. Flowers scented. Calyx ± divided, lobes ovate, 6–10 × 1.5–3 mm, spreading or reflexed at anthesis, stellate-hairy outside. Petals cream to white, ovate, 5–9 × 2–3 mm, glabrous. Stamens in 5 fascicles of 2, free portion of filaments 0.5–1 mm long; anthers 2–4 mm long; staminodes white to cream-coloured, 5, 4–6 mm long, papery. Ovary 1–2 mm long, 1.5 mm in diameter; style 0.5–1 mm; stigma lobes 1–1.5 mm long, cream-coloured, drying black. Capsule obconic, costulate towards the base, 12–25 mm long, 15–25 mm wide, apex concave with raised rim with expanded horns; seeds 1 per locule, winged, ovate, 3–7 × 2–4 mm, wings 8–10 × 4–6 mm.

KENYA. Kilifi District: near Roka, June 1937, *Dale* 1127! & Mwarakaya, 13 Dec. 1990, *Luke & Robertson* 2627!; Kwale District: Jombo Hill, 5 May 1985, *Barnett, Dorr & Cheek* 518!
TANZANIA. Tanga District: 8 km SE of Ngomeni, 2 Aug. 1953, *Drummond & Hemsley* 3602!; Uzaramo District: Pande Forest Reserve, 23 Apr. 1970, *B.J. Harris* 4467! & Pugu Hills, near mine, 27 Aug. 1982, *Hawthorne* 1589!
DISTR. **K** 7; **T** 3, 6; Mozambique
HAB. Evergreen coastal forest; 1–500 m

SYN. *Cistanthera holtzii* Engl. in E.J. 39: 578 (1907); Engl., V.E. 3(2): 352, t. 170 (1921); T.T.C.L.: 593 (1949)
    C. parvifolia Dale in T.S.K.: 41 (1936), *nom. nud.*
    C. parvifolia Milne-Redh. in K.B. 1937: 411 (1937). Type: Kenya, Kilifi District: Arabuko, *Graham* 1994 (K!, holo.; BR!, EA!, K!, P!, iso.)
    Nesogordonia holtzii (Engl.) Capuron in Not. Syst., Paris 14: 259 (1952), *comb. illeg.*; K.T.S.L.: 166, fig. (1994)
    N. parvifolia (Milne-Redh.) Wild in F.Z. 1(2): 551, t. 103, fig. B (1961); Capuron in Not. Syst. 14(4): 259 (1953), *comb. illeg.*; K.T.S.: 549 (1961); Wild & Gonç. in F.M. 27: 36, 37, t. 5, fig. E (1979); Vollesen in Opera Bot. 59: 34 (1980)

LOCAL USES. The timber is hard and heavy, but not durable. It is used in house building (Dorr pers. obs.)
CONSERVATION NOTES. This species appears to be widespread and fairly common on the basis of specimen sites alone (± 20 fide Barnett op. cit.). However its evergreen forest coastal habitat in which it can be dominant or co-dominant, has seen marked reduction in area in recent decades and is significantly threatened by touristic-linked development such as hotel building in Kenya. For the moment the species is rated here as near threatened but when quantitative data are available it is likely to rate threatened under category A of IUCN (2001).
NOTE. Vollesen, in his Selous checklist, records this taxon from **T** 8, but we have not seen material. The confusing nomenclatural history of this taxon, including the illegality of the Capuron combinations, is explained in Dorr & Barnett loc.cit.

2. **Nesogordonia kabingaensis** (*K.Schum.*) *Germain* in F.C.B. 10: 225, fig. 6C (1963); Capuron in Not. Syst., Paris 14: 259 (1953), *comb. illeg.*; Hamilton, Uganda For. Trees: 120 (1981). Type: Congo-Kinshasa, Kasai to Kabina, Sankuru R., *Laurent* s.n. (BR!, holo.)

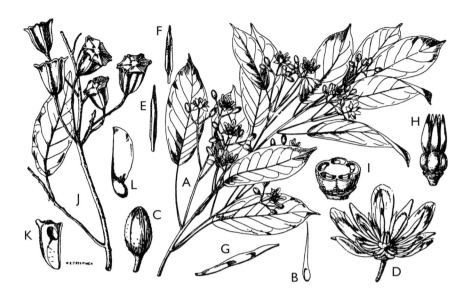

FIG. 16. *NESOGORDONIA KABINGAENSIS* — **A**, habit × $\frac{1}{4}$; **B**, stipule × ?; **C**, flower bud × 1.5; **D**, open flower × 1; **E** & **F**, stamens × 3; **G**, staminode × 4; **H**, ovary × $\frac{2}{3}$; **I**, section of ovary × 12; **J**, fruiting branch; **K**, fruit segment × $\frac{1}{3}$; **L**, seed × 1. Reproduced from FWTA ed. 2, 1: 313, fig. 119. Drawn by W.E. Trevithick.

Tree, 15–50 m tall; bole straight, narrow, one-half to two-thirds the height of a tree, slightly flaring to sharply buttressed at the base; bark variable in colour from whitish, grey, dark olive, to greyish-brown, fissured and flaking; crown smallish and rounded to pyramidal young branchlets tan to reddish-brown to black, smooth or striate, pubescent with stellate hairs, becoming glabrous in age. Leaves elliptic to obovate, 7–14.5 cm long, 3.2–6.8 cm wide; apex acuminate and mucronate, margin entire to sinuate toward to apex and slightly revolute, base obtuse/rounded or rarely cuneate; glabrous below; glabrous and glossy with a few scattered stellate hairs along the midvein above; below with domatia in vein axils; petiole 1.3–4.2 cm long, finely stellate-pubescent; stipules triangulate, fugaceous. Inflorescences 2–3-flowered corymbs (sometimes appearing 1-flowered by abscission); peduncle and pedicels densely stellate-pubescent; peduncle 2–3.5 cm long; pedicels 0.8–1.5 cm long, articulated 2–3 mm below the flower. Floral buds ovoid, 7–9 mm long, 5–7 mm wide. Flowers sweet-scented. Calyx lobes lanceolate to ovate, 8–12 mm long, 3–4 mm wide, densely stellate-pubescent outside. Petals broadly obovate, 9–10 mm long, 6–7 mm wide, white or cream-coloured, glabrous. Stamens in 5 fascicles of 3 each, consisting of 1 external and 2 internal stamens, free portion of filaments ± 0.5–1 mm long, anthers 2.5–5 mm long; staminodes 5, 6–7 mm long, coriaceous. Ovary ± 1 mm long, ± 1 mm in diameter; stigma lobes 1–2 mm long, + sessile. Capsule obconic, costulate toward the base, 3.1–3.5 cm long, 2.3–2.6 cm in diameter, densely covered with minute, golden-brown hairs, carpels flaring at the apex, summit prolonged into an erect horn, and apex deeply concave; inner carpel walls with stellate hairs in the seed cavities; seeds ± 7 mm long, 3–4 mm in diameter, seed wings 12–13 mm long, 6–7 mm wide. Fig. 16.

UGANDA. Toro District: Semliki Forest, Oct. 1905, *Dawe* 639! & Bwamba, between Bundibugyo and Butogo, Oct. 1940, *Eggeling* 4058! & Kitengya, Aug. 1937, *Eggeling* 3366!
DISTR. **U** 2; a widespread and variable species of W Africa and the Congo Basin
HAB. Riparian forest; ± 750 m

SYN. *Cistanthera kabingaensis* K.Schum. in E. & P. Pf., Nachtr. zum II-IV: 234 (1898); Th. & H. Durand, Syll.: 74 (1909); I.T.U., ed. 2: 414 (1952)

C. dewevrei De Wild. & Th.Durand in B.S.B.B. 38(2): 174 (1899); Th. & H. Durand, Syll.: 74 (1909). Type: Congo-Kinshasa, Monsembe, *Dewevre* 857 (BR, holo.)

C. ituriensis De Wild., Pl. Bequaert. 1: 478 (1922). Type: Congo-Kinshasa, Penghe, Ituri River, *Bequaert* 2448 (BR, holo!, iso!)

C. leplaei Vermoesen, Man. Ess. For. Congo: 49, t. p. 48 (1923). Type: Congo-Kinshasa, Mayumbe (Temvo), *Vermoesen* 1712 (BR!, holo., BR!, G!, iso.)

Nesogordonia dewevrei (De Wild. & Th.Durand) Capuron in Not. Syst., Paris 14: 259 (1953); Germain in F.C.B. 10: 224, fig. 6B (1963)

Nesogordonia leplaei (Vermoesen) Capuron in Not. Syst., Paris 14: 259, fig. 6A (1953), *comb. illeg.*

CONSERVATION NOTES. This species appears to be very widespread and often very common, and its main African habitat, semi-deciduous forest, although much diminished in West Africa, is still intact, as far as is known, in the Congo basin. It is rated here as of near threatened for conservation on the basis of reports that the Congo basin is to be made accessible for the export of timber.

12. **BYTTNERIA**

Loefl., Iter. Hisp.: 313–314 (1758), *nom. cons.*; Cristóbal in Bonplandia 4: 1–428 (1976); Bayer & Kubitzki in Fam. Gen. Vasc. Pls. 5: 244–245 (2003)

Lianas, shrubs, or small trees; stems unarmed (FTEA) or armed with prickles or spines; pubescent with stellate or simple hairs, or both intermixed, sometimes also with minute glandular hairs. Leaves simple (FTEA) or lobed, palmately nerved from the base, pinnately nerved above, foliar nectary normally one per leaf on the midvein below and near the base of the blade (FTEA), or 3–5 on one each of the basal nerves, uni- or multi-aperturate (FTEA); stipulate, the stipules generally fugaceous. Inflorescences axillary, 3–many-flowered cymes; erect or pendulous. Flowers bisexual. Calyx 5-parted, sepals almost free. Petals 5, free, unguiculate, cucullate, often with conspicuous lateral wings, and a distal appendage (lamina), appendage generally fleshy, simple; staminal tube campanulate or urceolate, stamens 5, antipetalous, alternating with 5 staminodes, anthers di-thecal, sessile, subsessile, or with noticeable filaments, longitudinally dehiscent, staminodes fleshy, sessile or not. Ovary syncarpous, sessile or subsessile, 5-locular, covered with incipient spines; 2 ovules per locule, ovules anatropous, superposed, the lower one aborting; style ± united to apex; stigmas apical, inconspicuous, 5-fid or subentire. Capsules ± globose, echinate, dehiscing septicidally and loculicidally, or only septicidally, into 5, single-seeded mericarps. Seed ovoid or elongate, testa smooth or tuberculate, sometimes strophiolate, exalbuminous; cotyledons foliaceous, bilobed, wrapped around the radicle; germination epigeal.

The genus includes ± 135 species, found in tropical and subtropical America, Asia, Polynesia; Madagascar (± 27 species); 6 species occur in tropical Africa.

The name of the genus has often been misspelled, e.g. as '*Buettneria*' or '*Buettnera*'

1. Lianas or scandent shrubs (rarely trees); leaves widely cordate basally, 5–7-nerved from the base, domatia lacking tufts of hairs; petals without wings; capsules densely spiny, spines acicular, 4–15 mm long .... 1. *B. catalpifolia*

   Shrubs or small trees; leaves cuneate or rounded to truncate basally, 3-nerved from the base, domatia with tufts of simple and fasciculate hairs; petals with recurved wings; capsules ± remotely spiny, spines stout, 2–4 mm long .... 2. *B. fruticosa*

1. **Byttneria catalpifolia** *Jacq.* in Hort. Schön. 1: 21–2, t. 46 (1797). Type not designated.

Liana or scandent shrub, rarely tree (Ethiopia), to 10(–15) m tall; pith soft, solid; young stems, leaves, and inflorescences subglabrous to puberulous or densely pubescent with small scale-like stellate hairs. Leaves ovate or ovate-lanceolate, 8–20(–24) cm long, 4–16 cm wide, apex acute to acuminate, margin entire (irregularly serrate in juvenile leaves), base cordate to truncate (sometimes oblique in juvenile leaves), sparsely pubescent above, and small stellate hairs common near petiole apex, sparsely pubescent below, with small stellate hairs especially on the midrib and veins (and simple hairs in juvenile leaves); palmately 5–7-nerved from the base, pinnately nerved above; domatia consisting of flaps of tissue without hairs in vein axils; foliar nectary at the base of the midrib beneath, lanceolate, multiaperturate; membranaceous; petiole 3–18(–23) cm long, puberulous like the young stems, constricted and slightly contorted at the base and apex when dry; stipules subulate to lanceolate, 4–5 mm long, puberulous with small stellate hairs, caducous, leaving a conspicuous scar. Inflorescences axillary cymes, 5–15 cm long; peduncle to 5 cm long, densely puberulous with stellate hairs; pedicels 7–12 mm long, articulated, puberulous; bracts and bracteoles minute, caducous. Floral buds long-turbinate. Sepals lanceolate-subulate, 5–7 mm long, ± 1 mm wide, puberulous with stellate hairs. Petals 5–7 mm long, the claw ± 1 mm long, geniculate, the hood ± 1 mm wide, without wings, the dorsal appendage (lamina) ± flattened, 5–6 mm long, glabrous. Staminal tube ± 1–1.5 mm long; anthers subsessile, ± 0.5 mm long, staminodes minute. Ovary ovoid, 0.5–1 mm long, 0.5–0.8 mm in diameter, muricate, hispid when young, becoming long-hispid; style 1–2 mm long, stigmatic surface 5-lobed. Capsules blackish-violet when mature, depressed globose, 2–3.5 cm long, 2.5–3 cm in diameter (excluding spines), surface of fruit glabrous, densely spiny, spines acicular, 4–15 mm long, dehiscing loculicidally, then incompletely septicidally the apex, the inside walls of each locule densely floccose with flattened, hyaline hairs; seed 1 per locule, reddish-brown, ovoid, 7–8 mm long, 4.5–6 mm wide, smooth, glabrous, sometimes maculated with lighter spots, raphe conspicuous. Fig. 17, p. 97.

subsp. **africana** (*Mast.*) *Exell & Mendonça* in C.F.A. 1: 197 (1951); F.W.T.A., ed. 2, 1(2): 314 (1958); Germain in F.C.B. 10: 219 (1963); Cristóbal in Bonplandia 4: 357 (1976); El Amin, Trees Shrubs Sudan: 107 (1990); Vollesen in Fl. Eth. 2(2): 166, fig. 80.1/1–9 (1995); Friis & Vollesen, Biol. Soc. Dan. Vid. Selsk. 51(1): 143 (1998). Type: "Lower Guinea, Congo", before 1859, *Chr. Smith* s.n. (K!, holo.)

Woody climber or scandent shrub. Leaves sparsely pubescent with very small, stellate hairs above and below, and simple hairs especially below when leaves are young, domatia in vein axils below inconspicuous flaps of tissue without hairs. Petal blade entire. Capsules rounded (not ribbed), 2–3.5 cm long, 2.5–3 cm wide (excluding spines), spines 4–15 mm long.

UGANDA. Bunyoro District: Budongo Forest, Nov. 1939, *Eggeling* 3835!; Toro District: Kabango, Bwambe, 22 Nov. 1935, *A.S. Thomas* 1514!; Mengo District: Mabira Forest, May 1908, *Ussher* 96!
DISTR. U 2, 4; Liberia, Ghana, Nigeria, Cameroon, Gabon, Central African Republic, Congo-Kinshasa, S Sudan (introduced?), Ethiopia, Angola
HAB. Forest and secondary growth; 900–1100 m

SYN. *B. africana* Mast. in F.T.A. 1: 239 (1868); Th. Durand & Schinz in Acad. Roy. Belg., Mém. 53(4): 81 (1896); K. Schum. in E.M. 5: 90, t. 5/B (1900); Th. & H. Durand, Syll.: 67 (1909); Staner in Ann. Soc. Sc. Brux., sér. 2, 58: 106 (1938); Exell in J.B. 65, Suppl.: 42–43 (1927)
    B. africana Mast. var. ? *angolensis* Hiern, Cat. Afr. Pl. Welw. 1: 92 (1896). Type: Angola, Golungo Alto, Sobati Quilombo-Quiacatubia near Muio R., *Welwitsch* 351 (BM!, lecto., chosen here (flowering material and leaves, sheet 000522462), BM!, isolecto. (labelled "Rio Mûio, Quilombo).

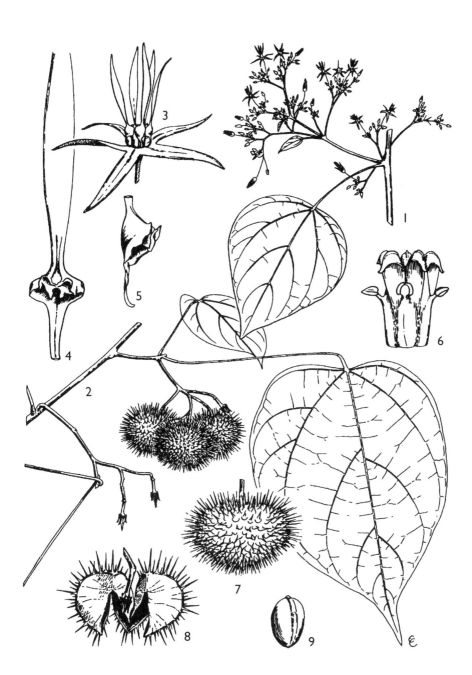

Fig. 17. *BYTTNERIA CATALPIFOLIA* — **1**, habit, in flower × ²⁄₃; **2**, habit, in fruit × ²⁄₃; **3**, flower × 6; **4**, petal, ventral view × 10; **5**, petal, basal part, lateral view, × 10; **6**, androecium × 20; **7**, fruit, immature × 1; **8**, fruit, dehisced × 1; **9**, seed × 2. 1, 3–6, from Breteler 1608; 2, 7& 9, from *Friis et al.* 3921; 8 from *de Wilde* 1231. Reproduced from F.E. 2(2): t. 80.1. Drawn by Eleanor Catherine.

CONSERVATION NOTES. This species appears to be very widespread and fairly common, and its habitat not significantly threatened as far as is known in the greater part of its range (central Africa). It is rated here as of "least concern" for conservation.

NOTES. The description in Vollesen (Fl. Eth. 2(2): 166 (1995)) is based, in part, on *Friis et al.* 4016, a sterile, juvenile specimen that exhibits characters such as oblique leaf bases and simple hairs that are not seen in any of the East African specimens examined. Cristóbal (Bonplandia 4: 348 (1976)), however, uses simple hairs below in juvenile leaves as one of the key characters to distinguish *B. catalpifolia* subsp. *africana* from the other subspecies.

The nominate subspecies is widespread in South and Central America.

2. **Byttneria fruticosa** *Engl.* in Abh. Königl. Akad. Wiss. Berlin 1894: 33 (1894); K. Schum. in P.O.A. C: 271 (1895) & in E.M. 5: 89–90, t. 5/A (1900); T.T.C.L.: 593 (1949); Cristóbal in Bonplandia 4: 341, tt. 5/A, 84/T (1976). Type: Tanzania, Lushoto District: Usambara, Gombelo, *Holst* 2159a (B†, holo.; K!, iso.)

Shrub or small tree, 1.5–5(–8) m tall; branches drooping. Leaves oblong, elliptic-oblong or broadly linear-oblong, 6–14(–28) cm long, 2–7(–12.5) cm wide (occasionally exceptionally large leaves are collected), apex acuminate to long-apiculate, margin entire to subundulate, base cuneate or rounded to truncate, glabrous above except for a few minute stellate hairs on the midrib, glabrous below except for scattered minute (almost scale-like) stellate hairs on the midvein and other large veins, palmately 3(–5)-nerved from the base, pinnately nerved above, domatia consisting of simple and fasciculate hairs and a small flap of tissue in the axils of the veins below, foliar nectary at the base of the midrib below, sometimes indistinct, multiaperturate; chartaceous; petiole 1–9 (16) cm long, ± pulvinate apically, sparsely stellate-pubescent, constricted and slightly contorted at the base when dry; stipules subulate, 5–7 mm long. Inflorescences pendulous, axillary cymes, 4–7 cm long; peduncle 2–3.5(–7) cm long; pedicels 0.4–1.5 cm long, articulated above the middle; peduncle and pedicels stellate-pubescent or sub-glabrous, both expanding in fruit. Floral buds narrowly turbinate, ± 2 mm long, with a stipe ± 1 mm long, densely stellate-pubescent or sub-glabrous. Sepals broadly lanceolate, 2.5–4 mm long, 1–1.3 mm wide, stellate-pubescent or sub-glabrous outside, always pubescent within. Petals white turning pink or pale yellow, 6–8 mm long, the claw ± 1.5 mm long, geniculate, tapering or more abruptly constricted, the hood and recurved wings 1.7–2 mm wide, the dorsal appendage (lamina) ± fusiform, 3–5.5 mm long, glabrous. Staminal tube ± 1 mm long, anthers sessile, 0.5–0.8 mm long; staminodes minute. Ovary turbinate, ± 1 mm long, ± 0.7 mm in diameter, distinctly papillate or warty; style 1–2 mm long, stigmatic surface 5-lobed. Capsules slightly ellipsoid, 1.6–2.3 cm long, 1.8–2.3 cm wide (excluding spines), surface of fruit with scattered stellate hairs or glabrous, spines widely-spaced, stout, straight or slightly unciform, (1–)2–4 mm long, dehiscing lonculicidally, then incompletely septicidally from the apex; seed narrowly elliptic, ± 12–15 mm long, ± 4–7 mm in diameter, glabrous, greyish-brown (sometimes glaucous), raphe conspicuous.

KENYA. Kilifi District: Pangani, crossing of Lwandani Stream on Chonyi–Ribe road, 17 Feb. 1977, *Faden et al.* 77/525!; Kwale District: Kaya Muhaka, 17 Sep. 1992, *Luke* 3303 & Shimba Hills, Buffalo Ridge, 15 Mar. 1991, *Luke & Robertson* 2706!

TANZANIA. Lushoto District: Mkomazi Game Reserve, Umba River by Umba Gate, 12 June 1996, *Abdallah et al.* 96/192!; Pangani District: S bank of Pangani River between Hale and Makinyumbe, 1 July 1953, *Drummond & Hemsley* 3126!; Tanga District: Kivomila River, 8 km SE of Ngomeni, 29 July 1953, *Drummond & Hemsley* 3529!

DISTR. **K** 7; **T** 3, 6; endemic

HAB. Moist forest, especially riverine habitats; 10–400 m

SYN. *Byttneria glabra* K.Schum. & Engl. in E.J. 39(3): 592 (1907); Engl. in V.E. 3(2): 444 (1921); T.T.C.L.: 593 (1949); Cristóbal in Bonplandia 4: 341 (1976), *non B. glabrata* K.Schum. Type: Tanzania, Pangani District, Useguha, Makinyumbi R., *Scheffler* 253 (B†, holo.) ["*B.* sp. nov." of K.T.S.L.: 161, fig (1994)]

CONSERVATION NOTES. Although this species appears to be fairly common in terms of numbers of specimen sites (± 20), its range is rather restricted and its habitat significantly threatened, accordingly it is here assessed as near threatened, provisionally.

NOTES. The type of *B. fruticosa* has glabrous flower buds. When there were very few collections of *Byttneria* from the East African coast, specimens with stellate-pubescent flower buds were separated out as a new species that was never published. The material is otherwise remarkably uniform and no character or combination of characters has been discovered that correlates with this minor pubescence variation. Plotting the distribution of a score of specimens of both forms confirms that there is no geographical separation either. Indeed, careful examination of the glabrous form often reveals the presence of at least a few stellate hairs. It therefore seems unlikely that we have anything but a single, somewhat variable species along the coast.

While this treatment was in press, Luke (pers. comm.) sent new records extending the geographical range to **T** 6 (*Luke & Luke* 4463 from Selous Game Reserve, Stiegler's Gorge), and extending the altitudinal range: the lowermost record given above being based on *Luke & Robertson* 2335 (Mwena River, Kwale District) and the higher on *Pakia & Luke* 1534a (E Usambaras near Sigi HQ).

One collection (*Luke & Robertson* 2706) has conspicuous galls on the under surface of the leaf blade.

13. **LEPTONYCHIA**

Turcz. in Bull. Soc. Imp. Naturalistes Moscou 31: 222 (1858); Germain in B.J.B.B. 31: 92–108 (1961); Hallé in Fl. Gabon 2: 129–133 (1961); Germain in F.C.B. 10: 226–240 (1963); Bayer & Kubitzki in Fam. Gen. Vasc. Pls. 5: 244 (2003)

Shrubs or small trees, sparsely stellate-pubescent. Leaves simple, entire, palmately nerved from the base, domatia in the axils of the principal veins below; stipulate. Inflorescences axillary, cymose, 1–many-flowered, or cymes on old wood; pedicels articulate; bracteate. Flowers bisexual, (3–)5-merous. Sepals free or almost so, sometimes unequal, rarely persistent in fruit, puberulous, usually with stellate hairs externally, stellate intermixed with simple hairs internally, and a ring of nectariferous tissue at the base. Petals scale-like, suborbicular, much shorter than the sepals, concave, without a distal appendage, margin ciliate, pubescent. Stamens 10 or 20 (rarely 12 or 18), filaments connate basally forming a short androecial tube, divided into 5 antipetalous phalanges of 2–4 stamens each and 0–4 long staminodes (resembling anther-less filaments), the phalanges alternating with 5 antisepalous short staminodes inserted on the interior of the androecial tube; short staminodes filiform, subulate or deltoid, sometimes scarcely distinct; anthers oblong, di-thecal, introrse, laterally dehiscent. Ovary syncarpous, sessile, (2–)3–5-locular; (2–)4 or more ovules per locule; style entire, slender, pubescent; stigma inconspicuous, briefly 2–5-lobed. Capsules globose to subglobose, woody to fleshy, pubescent (FTEA) or not, 2–3-locular or 1-locular by abortion, septicidal, loculicidal or opening irregularly. Seeds black or brown, shiny, glabrous, arillate, the fleshy aril thin, reddish or orange (yellow), not completely enveloping the seed, albuminous; embryo straight; cotyledons flat, thin.

At most 35–40 species in tropical Africa and southeast Asia; 12 narrowly circumscribed species (14 taxa) are found in the Congo-Kinshasa, whilst only 3 widely defined species (4 taxa) are found in southeast Asia.

A number of the African species of *Leptonychia* are narrowly circumscribed with distinctions being made on characters that were not considered to be reliable in a revision of the Asian species (Veldkamp & Flipphi in Blumea 32: 443–457, 1987). Accurate determinations of specimens are impossible without flowers, but even then infraspecific variation in androecial characters, in particular, contributes to the problem of delimiting species. Leaf shape also is variable and in some species groups it is exceedingly difficult to draw neat lines between currently recognized taxa. Veldkamp & Flipphi concluded that *Leptonychia* in Asia was over-described and the same may be true of the genus in Africa. A world-wide revision of the genus is in order. Intensive studies of character variation at the population level coupled with molecular analyses hold the most promise for understanding this genus.

In keeping with the treatment in F.C.B., a narrow (and traditional) circumscription of the East African species of *Leptonychia* is presented here.

Generic relationships have recently been clarified. Molecular data (Whitlock et al., Syst. Bot. 26: 420–437, 2001) show that *Leptonychia* belongs in the Byttnerieae where it appears to be sister to the tropical African *Scaphopetalum* Mast.

1. Filament-like long staminodes equalling or exceeding the
   fertile stamens in length 2
   Filament-like long staminodes shorter than the fertile
   stamens in length; **U** 2, 4 .... 1. *L. mildbraedii*
2. Shrubs, 1–3 m tall; leaf apex long-acuminate (leaf tips 3–4 cm
   long); sparsely pubescent with a ± uniform distribution of
   minute stellate hairs on the leaf surface below; **U** ? ...... 2. *L. semlikiensis*
   Shrubs or trees, 3–20 m tall; leaf apex acute to acuminate
   (leaf tips less than 3 cm long); very few minute stellate
   hairs on the leaf surface below and these scattered; **K** 4,
   7; **T** 3, 6, 7 .... 3. *L. usambarensis*

1. **Leptonychia mildbraedii** *Engl.* in E.J. 45: 322 (1911) & in Mildbr., Z.A.E.: 503 (1912); I.T.U., ed. 2: 421 (1952); Germain in F.C.B. 10: 235 (1963). Lectotype (as "holotype"), selected by Germain (1963: 236): Congo-Kinshasa, Kwidjwi Island in Lake Kivu, *Mildbraed* 1205 (B†, lecto.) [Note: Typification of this name is not complete and Germain's actions must be considered the first stage in the process of typifying the name since he did not cite a herbarium where the "holotype" was deposited and he stated explicitly that he did not see any of the four syntypes cited in Mildbraed's protologue.]

Shrub or small tree, to 12 m tall; crown sparse but spreading; bark smooth, grey or red-brown; young branchlets brownish-red, glabrous. Leaves elliptic to oblong-elliptic, 7–12.5(–15) cm long, 2.7–5 cm wide, apex acuminate, the tip 1–1.5 cm long, margin entire to imperceptibly repand, base obtuse to rounded or slightly cuneate, sometimes slightly oblique, deep brown and glabrous above, light brown and glabrescent below with scattered stellate hairs on the blade and multi-rayed hairs on the principal veins; palmately 3–5-nerved from the base with 4–6 pairs of ascending 2° veins; domatia in vein axils composed of ovate pits, mostly the margin glabrous but sometimes ringed with a few simple hairs; petiole 4–10 mm long, sparsely stellate-puberulous to glabrous; stipules 1–2 mm long, puberulous, abruptly caducous. Cymes sessile, ± stellate-puberulous, composed of 1–2(–3), 2–3-flowered dichasia; peduncle 2–4 mm long; pedicels 7–9 mm long, stellate-puberulous; bracts minute. Flowers whitish-green. Sepals linear-oblong, (6–)7–9 mm long, 1–2 mm wide, ± stellate-puberulous outside, densely puberulous inside. Petals suborbicular, apex acute, 1.2–1.5 mm long, 1.2–1.5 mm wide, puberulous or glabrescent externally, margin ciliate, glabrous internally. Staminal tube 0.7–1 mm long with 5 phalanges of 2 stamens (10 total) and 1–2 filament-like long staminodes (5–10 total) to 3 mm long, phalanges alternating with 5 short staminodes to ± 0.5 mm long; fertile stamens with free filaments 2–3(–4) mm long; anthers 0.6–1 mm long. Ovary 1.5–2 mm long, ± 2 mm in diameter, hirsute, 3(–4)-locular; 3 ovules per locule; style 3–6 mm long, hirsute almost to the apex. Capsules ± 2.5 cm long, 2–2.5 cm in diameter, brownish grey, papillate, densely tomentellous, loculicidal; seeds 1(–2) per locule, 10–12 mm long, 4–6 mm in diameter, brown, smooth, lustrous.

UGANDA. Toro District: Kibale Forest, near E boundary of forest immediately N of Kampala road, fl. 23 Feb. 1955, *Dawkins* D 836! & Kibale Forest, bank of river Mpanga, fr. 16 July 1938, *A. S. Thomas* 2303!; Mengo District: Kyagwe County, near Nansagazi, Nakiza Forest, fl. 18 Dec. 1950, *Dawkins* D 680!
DISTR. **U** 2, 4; **T** 4?; Congo-Kinshasa, and possibly Central African Republic

HAB. Moist forest and riverine forest, also in forest remnants; 1150–1800 m

SYN. [*L. multiflora* of De Wild., Pl. Bequaert. 1: 516 (1922), pro parte [as to *Bequaert* 3277]; I.T.U., ed. 2: 421 (1952); F.P.N.A. 1: 609 (1948), pro parte [as to *Bequaert* 3277], *non* K.Schum.
  [*L.* sp. of I.T.U.: 237 (1940)]

LOCAL USES. There is a report (Furuichi & Hashimoto, Pan Africa News 7(2): web version (2000)) of chimpanzees in the Kalinzu Forest, Uganda using bent branches of *L. mildbraedii* (and of other species) to make their ground beds.

CONSERVATION NOTES. This species is fairly widespread although not common. About 15 sites are known. In most of the range there are currently few threats to its habitat. Accordingly the species is here assessed as near threatened.

NOTES. *Kahurananga et al.* 2630 from **T** 4 may well belong to this species but has not been identified unequivocably.

  L. melanocarpa R.Germ., described from Congo-Kinshasa and known from the region between Lakes Kivu and Edward, might be encountered in Uganda. It can be distinguished from *L. mildbraedii* by its larger, black fruit; narrower leaves; and longer acuminate leaf apex.

2. **Leptonychia semlikiensis** *Engl.* in E.J. 45: 324–325 (1911), as *semlikensis*; Engl. in Z.A.E.: 504 (1912), as *semlikensis*; Germain in F.C.B. 10: 231 (1963). Type: Congo-Kinshasa, NW of Beni at Kwa Muera, *Mildbraed* 2140 (B†, holo.)

Shrub, 1–3 m tall; young branchlets reddish- to violet-brown, stellate-puberulous, glabrescent in age. Leaves elliptic-obovate to elliptic-ovate, 9–13 cm long, 2.8–3.7 cm wide, usually widest above the middle, apex long-acuminate, the tips 3–4 cm long, margin entire to repand, base rounded to slightly cuneate, greenish-brown to black and glabrous above, reddish-brown below with sparse, short stellate hairs and sometimes also long simple hairs along the midrib; palmately 3–5-nerved from the base with 6–8 pairs of ascending 2° veins; domatia in vein axils, widely orbicular, the opening defined by a few simple hairs, the pit also lined with scattered simple hairs; petiole 4–7(–9) mm long, ± pulvinate apically, stellate-puberulous; stipules lanceolate to acicular, 2–6 mm long, ± stellate-puberulous, caducous. Cymes subsessile, somewhat densely stellate-puberulous, composed of one or two 2–3-flowered dichasia; peduncle 2–3 mm long, pedicels 3–4 mm long (6–10 mm in fruit); bracts 1.5–2 mm long. Flowers greenish. Sepals ± lanceolate, 10–12 mm long, 1–1.5 mm wide, sparsely puberulous externally, puberulous basally and along the margin internally. Petals suborbicular, to 1.5 mm long, sparsely puberulous externally, glabrous internally. Staminal tube to 1 mm long with 5 phalanges of 2 stamens each (10 total) and (2–)3–4 filament-like long staminodes ((10–)15–20 total) equalling or exceeding the fertile stamens in length; phalanges alternating with 5 subtriangular to lanceolate-cuspidate short staminodes, 0.3–0.6 mm long; fertile stamens with filaments to (5–)8 mm long; anthers ± 1 mm long. Ovary to 2.5 mm long, hirsute, 3(–4)-locular; styles to 7–8 mm long, hirsute to the apex. Capsules reddish-brown, ± 1.3 cm in diameter, muricate, densely stellate-pubescent, loculicidal; seeds 2 per locule, black, to 10 × 6 mm.

UGANDA. No material seen (see Note)
DISTR. **U**; Congo-Kinshasa
HAB. Elsewhere in rain- and gallery forest, sometimes in swamps

SYN. [*L. multiflora* sensu De Wild., Pl. Bequaert. 1: 516 (1922), pro parte [as to *Bequaert* 3211]; F.P.N.A. 1: 609 (1948); I.T.U.: 421 (1952), pro parte [as to *Bequaert* 3211], *non* K.Schum.]
  [*Grewia batangensis* sensu De Wild., Pl. Bequaert. 1: 480 (1922), *non* C.H.Wright]

CONSERVATION NOTES. This species is known from only six sites in one province, Lacs Eduouard et Kivu, of Congo-Kinshasa, apart from its report in Uganda (Germain op.cit. 232). Forest clearance in this province has been extensive (Cheek pers. obs.) Accordingly the species is here assessed as VU B2a, b(iii), i.e. vulnerable.

Notes. The presence of this species in FTEA is based solely on the report in Germain (1963), who stated simply "Uganda". Both *L. semlikiensis* and *L. mildbraedii* have been confused with *L. multiflora* K.Schum., which does not occur in FTEA and apparently is restricted to Cameroon, Congo-Kinshasa, and possibly Nigeria (Germain, 1963; Pauwels in Scripta Bot. Belg. 4: 210, t. 225 (1993)). *L. multiflora* can be distinguished from the species found in FTEA by its 4–5-locular ovary.

3. **Leptonychia usambarensis** *K.Schum.* in E.J. 33: 313 (1904); T.T.C.L.: 600 (1949); K.T.S.: 549 (1961); Dows.-Lem. & F. White in B.J.B.B. 60: 84 (1990); K.T.S.L.: 166, fig. (1994); Schulman et al., Trees Amani Nat. Res.: 150–151 (1998); F. White et al., Evergreen For. Fl. Malawi: 562, fig. 197, D, E (2001). Types: Tanzania, Lushoto District: Usambara, Derema, *Scheffler* 155 (B†, syn.) & Kwai, *Albers* 301 (B†, syn.)

Shrub or trees, 3–20 m tall, 30–40 cm DBH; bark grey to brown, smooth or rough with numerous lenticels, slash light yellowish-brown, rapidly turning dark rusty brown; young branchlets minutely stellate-puberulous. Leaves oblong to elliptic or obovate-oblong, 4–22 cm long, 1.5–9 cm wide, apex acute to acuminate, margin entire to repand, base rounded or widely cuneate, glabrous above, ± glabrous below except for scattered appressed stellate hairs on the blade and scattered appressed multi-rayed hairs on midvein and major veins below, shiny dark green and nitid above, somewhat paler and glaucous (especially on nerves) below; palmately 3–5-nerved from the base with 4–6 pairs of ascending 2° nerves; domatia in vein axils composed of broadly ovate to orbicular pits ringed by simple hairs; petiole 5–15 mm long, ± pulvinate apically, puberulous to glabrescent; stipules acicular, to 4 mm long, puberulous, caducous. Cymes few-flowered; peduncle 6–8 mm long (to 15 mm in fruit); pedicels 6–15 mm long, stellate-pubescent, articulate; bracts to 2 mm long. Flowers white fading to yellow, whitish-green or yellow-green. Sepals greenish-white, subulate, sometimes unequal in length, 7–13 mm long, 2–3 mm wide, scattered stellate hairs externally, puberulous internally. Petals obcordate to rotund, 1.2–1.5 × 1.2–1.8 mm, apex ± truncate, ciliolate, a few stellate hairs externally, ± glabrous internally. Staminal tube 0.7–1.3 mm long with 5 phalanges of 2 stamens each (10 total) and 3–4 filament-like long staminodes (15–20 total) exceeding the fertile stamens in length, long staminodes crispate (curled), to 6.8 mm long; phalanges alternating with 5 gland-like short staminodes, ± 0.5 mm long; fertile stamens with filaments to 4–5.3 mm long, filaments erect; anthers ± 0.7 mm long; occasionally deviating specimens with 20 stamens total and no filament-like staminodia. Ovary 1.8 × 1.2 mm, hirsute, 3(–4)-locular, (3–)4–5 ovules per locule; styles ± 5 mm long, pubescent to the apex; stigmas barely lobed. Capsules yellowish-green at maturity, globose, asymmetrical, 10–17 mm long, 8–15 mm in diameter, shortly apiculate, papillate or muricate, densely stellate-pubescent, becoming glabrescent in age, loculicidal; seeds black, 1 per locule, 9–13 mm long, 7–8 mm in diameter, smooth, lustrous. Fig. 18, p. 103.

Kenya. Teita District: Chawia forest, fr. May 1985, *Beentje et al.* 868! & Ngangao Forest, 9 May 1985, *Faden et al.* 236!; Kwale District: Shimba Hills, Lango ya Mwagandi [Langomagandi], fl, fr 17 Mar. 1991, *Luke & Robertson* 2721!
Tanzania. Pare District: Chome Forest Reserve, fr. 27 Feb. 2001, *Mlangwa et al.* 1410!; Tanga District: Kwamgumi Forest Reserve, fr 13 June 2000, *Mwangoka* 1372!; Iringa District: Mwanihana Forest Reserve, fl. 8 Sep. 1984, *D.W. Thomas* 3657!
Distr. **K** 4, 7; **T** 3, 6, 7; ?Mozambique (*Luke et al.* 9915, n.v.); Malawi (Misuku Hills)
Hab. Moist forest; 200–350(–1900) m

Syn. [*L. usambarica* Engl. in N.B.G.B. 3: 84 (1900), *sphalm. pro* "*usambarensis*"]
    [*L. schliebenii* Mildbr., *nom. nud., in sched.*, based on *Schlieben* 2860]

Local uses. The bark of *L. usambarensis* is used for tying (*Semsei* 806).
Conservation notes. This species is relatively common (20 or more specimen-sites) and fairly widespread. However it is likely that it will rate as threatened under criterion A when quantitative data are available for its recent loss of habitat, so here it is asssessed as near threatened.

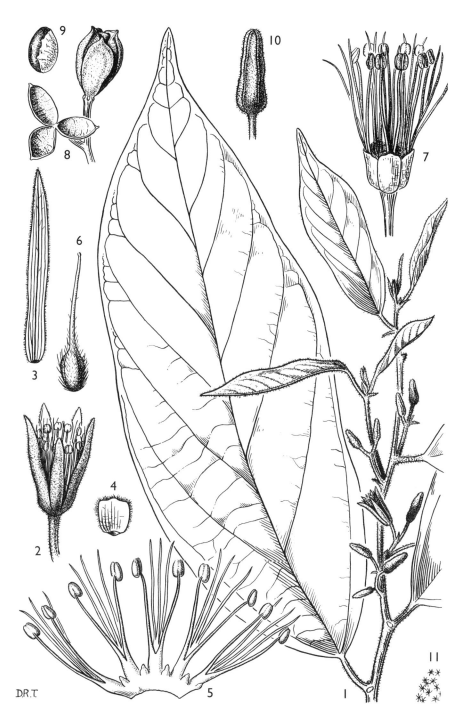

FIG. 18. *LEPTONYCHIA USAMBARENSIS* — **1**, habit × 1; **2**, flower × 3; **3**, sepal, inner view × 4; **4**, petal, inner view × 8; **5**, androecium, opened and spread, inside view × 5; **6**, ovary × 10; **7**, flower, with sepals removed × 6; **8**, fruits, with and without seeds × 1.5; **9**, seed, side view × 1; **10**, flower bud × 2; **11**, detail of stellate indumentum. Drawn by D.R. Thompson.

Cordeiro and Howe (Proc. Natl. Acad. U.S.A. 100: 14052–14056 (2003)) documented the negative impact of forest fragmentation on *L. usambarensis*, which is dependent upon birds for seed dispersal. Forest fragmentation in the mountains of East Africa is likely to contribute to local extinction of this (and other) tree species.

NOTES. Vegetatively, *L. usambarensis* and *L. mildbraedii* are very difficult to separate since they share the same vestiture, hair types, and domatia. The leaves of the former tend to be larger and less frequently acuminate than those of the latter, but there is overlap in these characters. Also, the pit domatia of *L. usambarensis* tend to be ringed by simple hairs and while those of *L. mildbraedii* for the most part are not, occasionally they are. In flower, most specimens of *L. usambarensis* and *L. mildbraedii* appear to be readily distinguishable by androecial characters (see key), but the fact that the Asian *L. caudata* (G.Don) Burret exhibits both of the staminodial types seen in these East African species, as well as intermediate types (see Veldkamp and Flipphi, 1987, fig. 3), tempers somewhat my enthusiasm for these key characters. In addition, a number of specimens of *L. usambarensis* have deviating androecial types. Notably *Mwangoka* 1771 and *Schlieben* 2860 lack filament-like staminodia and appear to have as many as 20 anthers, the anther sacs on those of the former are without pollen but those of the latter have pollen. Other specimens have anomalous flower buds: *Beentje et al.* 868 has mature fruit but also flower buds with short, fleshy sepals, reduced petals, 10 sessile anthers, and no visible ovary, ovules or style. Specimens with similarly shaped flower buds (e.g., *Mwangoka* 1112) show evidence of being parasitized by insects. All of this forces me to conclude that *L. usambarensis* and *L. mildbraedii* are closely-related vicariant species. Found in similar, higher elevation habitats, populations of these two species are now geographically isolated and presumably will continue to diverge.

14. HERMANNIA

L., Sp.Pl. 2: 673 (1753)

Mahernia L.

Herbs or shrubs, usually stellately hairy, sometimes also with simple or glandular hairs. Leaves simple, penninerved; petiolate; stipules foliaceous to subulate, persistent or early caducuous. Inflorescence axillary, 1-flowered or a terminal thyrse, rarely a simple raceme, usually bracteate. Flowers pedicellate. Calyx united at the base, 5-lobed, rarely inflated. Petals opposite the stamens, often clawed, the blade often reflexing at the junction with the claw, sometimes asymmetrical; claw often with margins involute, rarely hairy. Androgynophore short and inconspicuous, sometimes absent. Stamen filaments ligulate, terete, or winged, usually obovate, rarely with transverse appendages and so cruciform, rarely connate; anthers subulate, forming a cone around the ovary from which the style is exserted, extrorse, with longitudinal slits, but apparently functionally porose, lip of slits with bristles, medifixed but not versatile, connective inversely Y-shaped. Ovary superior, rarely stipitate, 5-locular, locules with 2 rows of numerous axile ovules, outer surface densely tomentose, apex sometimes awned, styles 5, ± united at anthesis, apex capitate or truncate. Fruit a loculicidally dehiscent capsule, rarely helically twisted, inflated and indehiscent, or a schizocarp with 5 several-seeded mericarps; locules 2–20 seeded. Seeds black, brown or dull red, reniform or subreniform, attatched by a short funicle on the smallest end; surface rugose or transversely banded, or foveolate, usually with minute papillae.

The description above and the key below are based solely on specimens from East Africa. When using the descriptions and key, bear in mind that the measurements of petals and stamens are made from hydrated material.

About 120 species in continental Africa, mostly in South Africa; three species in Arabia and four in Central & N America.

1. Annual herbs; inflorescences axillary, 1-flowered; flowers mauve, red, pink or yellowish green, rarely white or orange; fruits awned; seeds transversely rugose or banded 2
   Perennial shrubs; inflorescence a terminal thyrse or raceme, several-many-flowered; flowers yellow or orange; fruits lacking awns; seeds foveolate 5
2. Stipules triangular, length: breadth 1:1 to 2:1, rarely 3:1; pedicels with 2–3 patent bracteoles at the base; sepals as long as, or longer than petals; petals with hairy patch on inner surface .... 4. *H. glanduligera* (p. 109)
   Stipules lanceolate to filiform, length: breadth 3:1 or more; pedicels without bracteoles; sepals shorter than petals; petals glabrous 3
3. Flowers mauve, red or pink, rarely white; petals conspicuous, 7–8 mm long; sepals with stellate or non-strigose hairs; stem mostly with a mixture of glandular and crinkled simple hairs 4
   Flowers yellowish green; petals inconspicuous, less than 4 mm long; sepals with strigose simple hairs, hair bases swollen, red; stem indumentum subarachnoid-stellate ...... 3. *H. tigreensis* (p. 108)
4. Leaves narrowly elliptic or lanceolate, 1.7–2.6 (–3.6) × 0.6–1(–1.8) cm, margin toothed; peduncle densely glandular; filaments with hairy margins .... 1. *H. kirkii* (p. 107)
   Leaf-blade linear-oblong, 1.6–3 × 0.1–0.4 (–0.6) cm, margin entire; peduncle with a few simple hairs; filaments glabrous ...... 2. *H. modesta* (p. 108)
5. Stipules subulate or filiform 6
   Stipules lanceolate or ovate 15
6. Inflorescence glandular-hairy; calyx (including base) completely reflexed at anthesis; petals reflexed from the base, exposing the filaments and forming a cone-like shape 7
   Inflorescence simple or stellate-hairy, not glandular (except rarely in *H. macrobotrys*, very inconspicuously in *H. volleseni*); calyx lobes sometimes reflexed, but never the base; petals, if reflexed then from mid-length, filaments not exposed, petals forming a bell or disc-shape 10
7. Leaf-blade 3.4–5(–7) × 1.2–4.4 cm; inflorescence length (6–)15–30(–45) cm; partial-peduncle 3–6(–8)-flowered 8
   Leaf-blade (0.9–)1.5–3.8(–4) × (0.5–)0.6–2.8 cm; inflorescence length 5.5–16(–26) cm; partial peduncle 1(–3)-flowered 9

8. Inflorescence densely stalked-glandular-hairy, with a few simple hairs; partial-inflorescences (8–)12–14(–25); lowest partial-inflorescence 2–7 mm long; fruit longer than wide; Northern Uganda and Kenya . . . . 8. *H. boranensis* (p. 113)

Inflorescence stellate-hairy, rarely with a few stalked glandular hairs; partial inflorescences 25–37(–53); lowest partial peduncle (10–)12–18(–28) mm; fruit wider than long; Tanzania . . . . 6. *H. macrobotrys* (p. 111)

9. Inflorescence 5.5–9.5(–17) cm long; bracts (2–)2.5–3.5 mm long; fruit cylindrical-ellipsoid, 10 × 5–6 mm long, erect to pendulous, apex pointing away from stem 13. *H. fischeri* (p. 118)

Inflorescence 13–16(–26) cm long; bracts (0.2–)1.5–2 mm long; fruit globose, 4–5 × 5–6 mm long, pendulous-recurved, apex pointing at stem . . . . 14. *H. pseudofischeri* (p. 119)

10. Flowers erect or patent at anthesis, the pedicel straight or slightly curved; partial peduncle with two or more flowers, inflorescence thyrsoid 11

Flowers nodding, the pedicel u-shaped; partial inflorescences 1-flowered, inflorescence racemose . . . . 11. *H. pseudathiensis* (p. 115)

11. Inflorescence 2–6 cm long (but 12 cm in one specimen of *H. waltherioides*), each with 3–7 partial-inflorescences, borne on numerous spur shoots as well as on main axis 12

Inflorescence 8–30(–45) cm long, each with 7–37 partial-inflorescences, borne usually only on the main axis and not on spur shoots 14

12. Staminal filament cruciform, i.e. with two opposite, conspicuous transverse appendages ± ²⁄₃ the length from the base . . . . 9. *H. waltherioides* (p. 113)

Staminal filament oblong, lacking any appendages 13

13. Flower 8–12 mm diameter; petals 6–8(–9) mm long; fruit helically twisted, inflated . . . . . . 12. *H. uhligii* (p. 117)

Flower 14–17 mm diameter; petals 10.5–12 mm long; fruit straight, not inflated . . . . 10. *H. athiensis* (p. 114)

14. Stipules 1.7–3 mm long; pedicellar bracts absent; petals 3.5–5.5 mm long; fruit oblate . 6. *H. macrobotrys* (p. 111)

Stipules 5–7 mm long; pedicellar bracts present, filiform; petals 5–6.5 mm long; fruit ovoid-ellipsoid . . . . 7. *H. oliveri* (p. 112)

15. Habit prostrate; stipules lanceolate; calyx inflated 16. *H. volleseni* (p. 121)

Habit erect; stipules ovate; calyx not inflated 16

16. Stipules broadly ovate, 7–10 × 5–7 mm; flowers (6–)12–15(–17) mm wide . . . . 5. *H. exappendiculata* (p. 110)

Stipules narrowly ovate, 2–4 × 1.5–2 mm; flowers 4–5 mm wide . . . . 15. *H. paniculata* (p. 119)

1. **Hermannia kirkii** *Mast.* in F.T.A. 1: 233 (1868); K. Schum. in P.O.A. C: 270 (1905) & E.M. 5: 84, t. 4E (1900); Wild in F.Z. 1: 547 (1961); Wild & Gonç. in Fl. Moçamb. 27: 34 (1979); Blundell, Wild Fl. E. Afr.: t. 736 (1987); U.K.W.F. ed. 2: 96 (1994); Vollesen in Fl. Eth. 2: 181 (1995). Type: Botswana, near Ghanzi, Oct.-Nov. 1861, *Baines* s.n. (K! lecto., selected here)

Erect annual herb 7–35(–60) cm tall, stem 1–2(–5) mm diameter at base, generally with 3–4 subequal ascending branches from the lower half; branches densely invested in a mixture of non-glandular simple crinkled and glandular simple hairs, sometimes scattered with longer simple hairs. Leaf-blade narrowly elliptic or lanceolate, 1.7–2.6(–3.6) × 0.6–1(–1.8) cm, apex acute, margin with 6–7(–14) teeth on each margin, base rounded, 3–4 nerves on each side of the midrib, upper surface with thinly scattered simple glandular and non-glandular hairs, lower surface with sessile glandular hairs and thinly scattered stellate hairs, sometimes admixed with appressed simple hairs; petiole 0.5–0.8(–1.1) cm long, indumentum as stem; stipules subulate to lanceolate, 1.2–1.5(–3.5) × 0.3(–1) mm. Inflorescence with flowers single in the upper axils; peduncle (8–)10–20 mm long, ebracteate, densely glandular-hairy; pedicel thicker than and inconspicuously articulated with the peduncle, 1–2 mm long. Flowers pendulous, red, mauve, pink or, rarely, orange, 6–8 mm wide. Calyx campanulate, 5–7 × 5–8 mm, lobed for $^1/_3$, lobes widely triangular, 2–3 × 1–2 mm, apex acuminate, 0.5–1 mm long, densely glandular-hairy. Petals oblanceolate to obovate, 7–8 × 3–5 mm, margin involute in the lower half, glabrous. Stamens with filaments oblanceolate, 3–4 × 1.2–1.3 mm, with simple hairs on the distal margin; anthers blue, 3.7–5 × 0.5–1 mm. Ovary truncate-ovoid, slightly awned, 1.2–2 × 0.7–1.4 mm, stipe 0.5–2 mm long; styles 3.5 mm long, puberulous. Fruit capsule truncate-ovoid, awned, 4–6(–7) × 6–7(–8) mm; awns widely triangular, subpatent, shortly glandular-hairy; seeds subreniform, 1–1.3 × 0.8–1 × 1 mm, widest at chalazal end, rugose, minutely papillate, dark brown.

UGANDA. Karamoja District: Upe, Lossom, 9 July 1956, *Dale* U878! & Kangole–Moroto road, July 1930, *Liebenberg* 138!
KENYA. Northern Frontier District: 11 km S of Laisamis, Moile Hill, 25 Nov. 1977, *Carter & Stannard* 722!; Turkana District: 40 km SW of Lodwar, 12 May 1953, *Padure* 147!; Baringo District: 2–3 km W of Kampi ya Samaki, 29 Oct. 1992, *Harvey et al.* 18!
TANZANIA. Shinyanga District: Udhe, Shagihiru to Sekenke, 24 Jan. 1936, *Burtt* 5511!; Kondoa District: Kondoa, Great North Road, 16 Jan. 1962, *Polhill & Paulo* 1192!; Iringa District: Iringa, Iringa College of National Education, 29 Jan. 1972, *Pedersen* 726B!
DISTR. **U** 1; **K** 1–3; **T** 1, 5, 7; Angola, Mozambique, Zimbabwe, Botswana, Namibia, South Africa
HAB. *Acacia-Commiphora* bushland, wooded grassland, or semi-desert, often at margin of seasonally wet areas; 250–1800 m

LOCAL USES. Eaten by goats, sheep, horses etc. (*Mwangangi & Gwynne* 1248)
CONSERVATION NOTES. This species is widespread and common, and its habitat not significantly threatened. It is rated here as of "least concern" for conservation.
NOTES. *H. kirkii* is closely related to *H. modesta* K.Schum. and indeed is treated by Verdoorn (Bothalia 13: 1–63, 1980) as a synonym (unjustifiably) of that species. *H. kirkii* is distinguished from *H. modesta* by the wider, toothed leaves (not linear, entire), proportionately shorter, densely glandular peduncle (versus longer, simple-hairy), hairy filaments and dense indumentum (versus glabrous filaments and sparse indumentum). *Hermannia kirkii* is often confused with *H. glanduligera* (q.v.).
　　I have lectotypified this name on the Baines specimen since this well represents the taxon we know today as *H. kirkii* and is annotated with this name by Masters. Amongst the other five syntypes are discordant elements.
　　A small annual cleistocarpic variant of this species occurs sporadically, especially at marginal parts of its range in East Africa. This variant has smaller, less showy flowers than the usual sort, all of which set fruit. It is not worthy of nomenclatural distinction.
　　Luke (pers. comm. while ms in press) extends the range to **K** 4 (*Faden* 68/172 ,Thika) and to **K** 7 (*Verdcourt* 891 Kora).

2. **Hermannia modesta** (*Ehrenb.*) *Mast.* in Oliv., F.T.A. 1: 232 (1868); K. Schum. in E.M. 5: 83, t. 4C (1900); Wild in F.Z. 1: 548 (1961); Vollesen in Fl. Eth. 2: 181 (1995); Thulin in Fl. Som. 2: 34 (2000). Type: Saudi Arabia, near Djedda, *Ehrenberg* s.n. (B, presumed destroyed)

Erect, annual herb 4–15 cm tall, stem 0.5–1 mm diameter at base, usually unbranched; stems scattered with patent simple and glandular hairs, sometimes with a few stellate and long simple hairs. Leaf-blade linear-oblong, 1.6–3 × 0.1–0.4(–0.6) cm, apex obtuse, margin entire, base acute, nerves inconspicuous, upper surface glabrous or with a few stellate or simple hairs, lower surface with scattered stellate hairs and sessile glandular hairs; petiole 1–2 mm long; stipules subpatent, subulate, 1.5–2.5 mm long. Inflorescence with flowers single in the upper axils; peduncle 12–30 mm long, ebracteate, subglabrous or with a few simple hairs; pedicel thicker than and conspicuously articulated with the peduncle, 2–3 mm long. Flowers pendulous, wine-red or red, ± 5 mm wide. Calyx campanulate, 4 × 3–4 mm, lobed for ½, lobes narrowly triangular, 2–2.6 × 1–1.3 mm, glabrous apart from a few simple hairs on the lobe margins. Petals oblanceolate, 5–6 × 1.4–3 mm, margins not or only very slightly involute in lower half, glabrous; stamens with filaments oblanceolate, 2–2.5 × 0.7 mm, glabrous; anthers white, 2.5–3 × 0.3 mm. Ovary truncate-ovoid, 1.5 × 1.2 mm, awns inconspicuous, stipe 0.7 mm long; styles 1.5 mm long. Fruit capsule truncate-ovoid, ± 5 × 4 mm, awns inconspicuous, 0.2–0.4 mm long, erect, sparsely and inconspicuously puberulous with simple and glandular hairs; seeds black, reniform, 1.3 × 0.7 × 0.6 mm, with ± 8 transverse bands and scattered glossy black papillae.

KENYA. Northern Frontier District: Lake Turkana [Rudolf], Koobi Fora to Shin, 1 May 1971, *Faden & Evans* 71/346! & 74 km S of Mado Gashi, 13 Dec. 1977, *Stannard & Gilbert* 1029!
DISTR. **K** 1; Sudan, Eritrea, Somalia, Angola, Zimbabwe, Botswana, Namibia and South Africa; Egypt and Arabian Peninsula
HAB. *Acacia-Commiphora* bushland or semi-desert scrub; 300–600 m

SYN. *Trichanthera modesta* Ehrenb. in Linnaea 4: 402 (1829)

LOCAL USES. None recorded.
CONSERVATION NOTES. Although known from only two specimens in the FTEA area, this species is widespread and common in its global range, and its habitat not significantly threatened. It is rated here as of "least concern" for conservation.
NOTES. Verdoorn (1980 *op.cit.*), appears to apply the name in South Africa in a wider sense than is used here.
    The absence of *H. modesta* from Uganda, Tanzania and Zambia gives it a disjunct distribution between SW and NE Africa.

3. **Hermannia tigreensis** *A.Rich.*, Tent. Fl. Abyss. 1: 74, t. 17 (1847); Mast. in F.T.A. 1: 233 (1868), as *tigrensis*; K. Schum. in E.M. 5: 85, t. 4B (1900) & in P.O.A. C: 270 (1905); Keay in F.W.T.A. 1: 318 (1958); Wild in F.Z. 1: 546 (1961); Germain in F.C.B. 10: 255 (1963); Wild & Gonç. in Fl. Mocamb. 27: 33 (1979); Vollesen in Fl. Eth. 2: 181 (1995). Types: Ethiopia, Djeladjeranne, *Quartin-Dillon & Petit* s.n. (not located, syn.); Djeladjeranne, *Schimper* III: 1470 (FT, K!, P, syn.); Tekeze River, *Schimper* II: 812 (B, FI-W, K!, P, UPS, syn.)

Annual herb or short-lived perennial, 8–30 cm tall, erect, at length slightly spreading, stem 1–3 mm diameter at base, branching, the branches themselves branching, at length forming a rounded mass; stems densely covered with subarachnoid stellate hairs. Leaf-blade ovate, 2.5–4 × 1–1.4(–1.7) cm, apex acute, margin serrate, base rounded, with 7–10 coarse teeth, nerves ± 3 on each side of the midrib, upper surface with scattered simple, erect hairs, lower surface with sessile glandular hairs and stellate hairs; petiole 1.5–3(–4.5) cm long; stipules patent, lanceolate, 2.5–5.5 × 0.3–1.2 mm. Inflorescence with flowers single in the upper axils; peduncle 19–28 mm long, ebracteate, with a few scattered simple hairs; pedicel thicker than and inconspicuously articulated with the peduncle, 2–3 mm long. Flowers erect, green or yellow-green, 2.5–4 mm wide. Calyx campanulate, 3.5–4 × 2.5 mm, lobed for $\frac{1}{2}$, lobes narrowly triangular, erect, 2–2.5 × 0.8–1.2 mm, translucent, with strigose simple hairs with red swollen bases arising from the nerves. Petals oblanceolate-ligulate to spathulate, 3.5–3.9 × 0.7–1 mm, slightly involute in the lower $\frac{1}{3}$, glabrous. Stamens with filaments obovate, 1.3–1.8 × 0.6–0.8 mm, glabrous; anthers 1.2–1.8 × 0.2–0.3 mm. Ovary shortly ellipsoid-truncate, 2.2–2.5 × 1.6–1.8 mm, apex with 5 lobes, stipe 0.1 mm long; styles 1.2 mm long. Fruit 4–5 × 5–6 mm, truncate, awns erect, 0.5–1.2 mm long, densely covered in stellate hairs; seeds reniform, 1.2 × 1 × 1 mm, rugose, densely papillate, papillae pale brown.

UGANDA. Busoga District: Lake Victoria, Lugala landing, 26 March 1953, *G.H. Wood* 673!; Teso District: Serere at Kyere, July 1926, *Maitland* 1299!
TANZANIA. Moshi District: Moshi, Aug. 1927, *Haarer* 590!; Chunya District: N Lupa Forest Reserve, 5 March 1963, *Boaler* 876!; Masasi District: Masasi, 23 April 1935, *Schlieben* 6350!
DISTR. **U** 3; **T** 1, 2, 4, 6–8; West Africa, Sudan, Ethiopia, Angola, Malawi, Mozambique, Zimbabwe
HAB. Open woodland or lake edges; 450–1300 m

LOCAL USES. None recorded.
CONSERVATION NOTES. *H. tigreensis* is widespread in Africa and common, and its habitat not significantly threatened. It is rated here as of "least concern" for conservation.
NOTES. *H. tigreensis* can be reliably be distinguished from the species of the genus that are annual herbs (or short-lived perennials), namely *H. kirkii*, *H. modesta* and *H. glanduligera* by its yellowish green, small (± 4 mm long) flowers (versus red, mauve pink or white, 5 mm or more long).

4. **Hermannia glanduligera** *K.Schum.* in Verh. Bot. Ver. Prov. Brandenb. 30: 232 (1880) & in E.M. 5: 57 (1900); Wild in F.Z. 1: 545 (1961); Wild & Gonç. in Fl. Mocamb. 27: 31 (1979). Type: Namibia "Olukonda, (Ondongo-Stamm) Amboland" (not located, B †?)

Annual herb to short-lived perennial 15–100(–200) cm tall, usually viscid, stem 3–7.5 mm diameter at base, generally with 4–5 subequal ascending and prostrate branches from near the base; branches invested densely with predominantly simple, glandular hairs, sometimes mixed with simple crinkled hairs and with sporadic stellate hairs, less usually almost entirely stellate-hairy. Leaf-blade linear-lanceolate, lanceolate, or linear-oblong, rarely ovate or elliptic, 9–25(–48) × (2.2–)3–14(–22) mm, apex acute, margin serrate, base obtuse, with (1–)2–9 teeth on each side, nerves 2–4 on each side of the midrib, upper surface with sessile glandular hairs scattered with stellate hairs, lower surface densely covered with simple glandular hairs and stellate hairs; petiole 1–8(–10) mm long; stipules triangular, 1.3–1.8 × 0.6–1 mm. Inflorescence with flowers single in the upper axils; peduncle 5.5–18 mm long, indumentum as stem; with 2–3 patent triangular bracts 0.5–1.2 × 0.3–0.6 mm at the junction with the pedicel; pedicel 1.5–2(–3) mm long. Flowers pendulous, red, mauve, pink, rarely white or pink with red centre, 5–8 mm wide. Calyx campanulate, 5–6 × 5 mm, lobed for $\frac{1}{2}$, lobes 3–4.1 × 1–1.4 mm, indumentum as stem. Petals obovate or subligulate, (4.2–)5–6 × (1.9–)2.5–3 mm, apex rounded, basal $\frac{1}{2}$ to $\frac{1}{3}$ ± involute, rarely so involute that petal appears clawed; band of simple hairs above involute zone. Stamens with filaments oblanceolate, 2–3.2 × (0.8–)1.2–1.8 mm, with or without hairs on the distal margin; anthers 2.5–5.5 × 0.5–1 mm. Ovary ellipsoid, 1.7–2 × 0.9–1.5 mm, truncate, slightly awned, stipe 0.2–0.7 mm; style 3.5–5 mm. Fruit capsule truncate-ovoid, awned, 5–7 × 5.5–8 mm, awns patent, 2(–4) mm long, indumentum as stem; seeds reniform, 1.2–1.5 × 1–1.2 × 1 mm, rugose, densely white papillate.

KENYA. Masai District: Nguruman Hills, Lenyora, 27 Sep. 1944, *Bally* 3835!; Kitui/Tana River District: 92 km on Garissa–Nairobi road, Katumba Hill, 14 May 1978, *Gilbert & Thulin* 1717!; Teita District: Ndi, Manga Hill [Simba], 12 April 1978, *Verdcourt* 5313!

TANZANIA. Mpanda District: Lake Katavi, 10 Feb. 1962, *Richards* 16054!; Kondoa District: Kinyasi, 1 Jan 1927, *B. Burtt* 917!; Kilosa District: 3 km N of Mbunyi, 30 Aug. 1970, *Thulin & Mhoro* 810!

DISTR. **K** 4, 6, 7; **T** 2, 4, 5–7; Angola, Zambia, Malawi, Mozambique, Zimbabwe, Botswana, Namibia, South Africa

HAB. Road edges, old cultivation, edges of seasonally wet areas and *Acacia* bushland or woodland, often on sandy or sandy clay soils; 30–1200(–1400) m

SYN. *H. teitensis* Engl. in E.J. 55: 370 (1919). Type: Kenya, Teita District: Ndara, *Hildebrandt* 2385 (B, holo., presumed destroyed, K! iso.)
    H. viscosa sensu Agnew, U.K.W.F. ed. 2: 96 (1994), *non* Hiern

LOCAL USES. Eaten by stock (*Hornby* 17).

CONSERVATION NOTES. This species is widespread and common, and its habitat either not significantly threatened or (old cultivation) increasing. It is rated here as of "least concern" for conservation.

NOTES. Most E African material of *H. glanduligera* has been misidentified as *H. kirkii* (see key for differences) or as *H. viscosa*, an Angolan species which lacks the hairy petals of *H. glanduligera*, as pointed out by Wild in FZ 1: 546 (1961).

    The indumentum of *H. glanduligera* is unusually variable in trichome complement for the genus and no assistance to identification. *H. glanduligera* can be easily recognized within the group of *Hermannia* in East Africa which are annuals, with single axillary flowers, obovate filaments and awned fruits (*H. kirkii*, *H. modesta* and *H. tigreensis*) by the widely triangular stipules, the presence of bracteoles at the base of the pedicel, the peduncles which are usually only as long or shorter than the leaves and by the petals, which are hairy on the inner surface.

5. **Hermannia exappendiculata** (*Mast.*) *K.Schum.* in Abh. Preuss. Akad. Wiss. 1894: 55 (1894); E. & P.Pf. III 6: 80 (1895); K. Schum. in E.M. 5: 51, 85, t. 4B (1900); P.O.A. C: 270 (1905); T.T.C.L.: 599 (1949); Blundell, Wild Fl. E. Afr.: t. 443 (1987); U.K.W.F. ed. 2: 97 (1994); Vollesen in Fl. Eth. 2: 180 f. 80.7, 4–5 (1995); Thulin in Fl. Som. 2: 34, t.18 D&E (2000). Type: Kenya, Mombasa, *Bouton* 1857 (K!, holo.)

Shrub 0.6–2(–2.5) m tall, stems red-brown, with white stellate hairs. Leaves ovate to lanceolate, 3.1–4.8(–8.5) × 1.6–3.1(–4.4) cm, apex rounded, margin dentate-crenate, base obtuse or rounded, rarely slightly cordate; nerves ± 7 on each side of the midrib, upper surface with scattered suberect simple or 2-armed hairs, sometimes stellate-hairy, lower surface densely covered with stellate hairs; petiole (7–)10–20 mm long; stipules leafy, asymmetric, ovate, 7–10 × 5–7 mm, apex acuminate, base unequally cordate, inserted obliquely on the stem, palmately 3–5-nerved, minutely stellate-hairy. Inflorescence a terminal thyrse 6–10(–20) cm long, partial-inflorescences 4–7(–9), 1.8–3.5(–7) cm long at base, each bearing (3–)5–10(–20) flowers, lowest partial peduncle (4–)7–15(–20) mm long; peduncular bracts filiform 3.5–4 mm long; partial-rhachis bracts 0.5–2 mm long; pedicellar bracts absent or single, filiform, patent, 0.5 mm long; pedicel 3.5–7 mm long. Flowers erect, yellow or orange, (6–)12–15(–17) mm wide. Calyx with the united basal part campanulate, 2.5–4.5 × 2.5–4 mm, the distal lobes spreading slightly, narrowly triangular, 3.5–6 × 1.2–1.6 mm, predominantly glandular-hairy, stellate-hairy, simple-hairy or a mixture of two of these. Petals asymmetric, clawed, sometimes weakly, 6–9 mm long, the blade elliptic, (2.7–)3.5–4 mm wide, forming the distal half to $^2/_3$, claw oblong, widening towards distal end, (0.4–)1.2–1.4 mm wide at the distal end, basal margins thickened, slightly involute, glabrous. Stamens with filaments subterete, 1.2–2 × 0.3–0.4, lacking appendages, glabrous; anthers 4–5 × 0.6–1. Ovary ellipsoid, 3 × 1.5, apex lacking awns; style 4.5–5.5 mm long. Fruit 5-angled, oblong in side-view, 11–13 × 8 mm, the valves obovate, the margin with a tuberculate ridge ± 0.5 mm high, sparsely stellate-hairy; seeds dull red, reniform, 1.5–1.6 × 1.2 × 1.2 mm, foveolate, scattered with glossy red papillae.

UGANDA. Karamoja District: eastern Mathemiko country, March 1959, *Wilson* 702!

KENYA. Northern Frontier District: Gurika Hill, 18 Nov. 1977, *Carter & Stannard* 516!; Machakos District: 68 km Mutome–Kibwezi, 23 Nov. 1979, *Gatheri et al.* 79/142!; Kilifi District: Mida, 1927, *Gardner* 1425!
TANZANIA Tanga District: Tanga, 3 Nov. 1929, *Greenway* 1837!; Bagamoyo District: Bagamoyo, May 1874, *Hildebrandt* 1259!; without locality, 1894, *Stuhlmann* s.n.; Pemba, Ole, 26 Oct. 1929, *Vaughan* 887!
DISTR. **U** 1; **K** 1–4, 7; **T** 3, 5, 6; **P**; Ethiopia, Somalia
HAB. *Acacia-Commiphora* bushland, roadsides, old cultivation, edge of sea, swamps, sometimes in coastal forest or grassland; 5–1900 m

SYN. *Mahernia exappendiculata* Mast., F.T.A. 1: 234 (1868)

LOCAL USES. None recorded.
CONSERVATION NOTES. This species is widespread and common, and its habitat not significantly threatened. It is rated here as of "least concern" for conservation.
NOTES. *Hermannia exappendiculata* is the most commonly collected species of the genus in East Africa. It is immediately recognizable by its large, leafy, asymmetric, broadly ovate stipules, unusual amongst the East African species of the genus. The only other species with ovate stipules is *H. paniculata*, but this has much smaller, narrowly ovate stipules and minute flowers and so is not likely to be confused with *H. exappendiculata*.
    H. exappendiculata is variable in petal size and particularly in calyx lobe length and indumentum. Calyx indumentum can be predominantly stellate-hairy, simple hairy or glandular-hairy. Specimens from **K** 1 (higher altitudes) tend to have the upper leaf surface simple hairy and larger flowers with pedicellate bracts, glandular-hairy calyces with longer lobes. Those from the coast (near sea-level) tend to have the upper leaf surface stellate-hairy, smaller flowers without pedicellate bracts and stellate-hairy calyces with shorter lobes. These distinctions are not always maintained. Although speciation may be in progress here, differences are not yet clear-cut enough to warrant recognition of infraspecific taxa.
    H. appendiculata is reported from Madagascar by Arenes (1959, Fl. Madag. 131: 138) on the basis of a single specimen *Bojer* s.n. for which the locality is unknown. This was probably collected in E Africa. No other specimen of this or any other species of *Hermannia* is reported from Madagascar.

6. **Hermannia macrobotrys** *K.Schum.* in E.M. 5: 51 (1900) & in E.M. 5: 85, t. 4B (1900); T.T.C.L.: 600 (1949). Type: Tanzania, District unclear, Rufiji, *Goetze* 81 (B, holo., probably destroyed; K!, iso.)

Shrub 0.3–0.9 m tall, stems lax, dark brown, stellate-hairy, rarely glandular-hairy. Leaves ovate or elliptic, 3–4.6(–7) × 1.5–2.4(–3.2) cm, apex rounded, obtuse or acute, margin crenate-serrate, base obtuse or rounded, nerves 5–7 on each side of the midrib, upper surface thinly stellate or simple hairy, lower surface concealed by white stellate hairs; petiole 0.7–1.7(–2.2) cm long; stipules narrowly triangular or linear-oblong, 1.7–3 × 0.5–1 mm, apex acute. Inflorescence a terminal thyrse (9–)15–24(–45) cm long, partial-inflorescences 25–37(–53), each 1.1–2.5 cm long at base, bearing 3–6(–11) flowers, lowest partial peduncle 2–7 mm long; peduncular bracts filiform, 2–5 mm long; partial-rhachis bracts 1.5–2 mm; pedicellar bracts absent; pedicel 1.3–3 mm long. Flowers erect, yellow or orange, 5–7 mm wide. Calyx with the basal part shallowly cup-shaped, 1–2 × 3–4.5 mm, lobes narrowly triangular, 3–5.5 × (0.8–)1.5–1.6 mm, stellate-hairy, densely so on inner surface. Petals revolute, obovate-spathulate, 3.5–5.5 × (1.3–)2–3 mm, margin not involute, glabrous. Stamens with filaments elliptic, flattened or winged, 6–10(–12) × 4–8(–10) mm, lacking appendages, glabrous; anthers 4–5 × 1–1.5 mm. Ovary globose, 1.5–2.5 mm diameter, apex lacking awns; style 3–5 mm long. Fruit capsule oblate, ± 8 × 10 mm sparsely stellate-hairy, with 5 longitudinal, echinate, densely stellate-hairy double ridges ± 0.5 mm high, marking the valve edges; seed subreniform, dull red, 1–1.4 × 0.8–1 × 1 mm, foveolate, minutely and sparsely papillate.

TANZANIA Mpwapwa District: Mpwapwa, 17 April 1932, *Burtt* 3867!; Kilosa District: Kimamba, near Kilosa, 24 Jan. 1926, *Burtt* 5101!; Iringa District: Mtera, 5 May 1971, *Mhoro* 1102!
DISTR. **T** 2, 5–8; Zimbabwe
HAB. Grassland, *Acacia-Commiphora* woodland or secondary vegetation; (50–)450–1400 m

LOCAL USES. None are recorded

CONSERVATION NOTES. This species is widespread in Tanzania although not particularly common (19 specimens are known, versus 90 for *H. exappendiculata*), and its habitat not significantly threatened. It is rated here as of "least concern" for conservation, although if a significant threat to its habitat occurs, it might be rated as "near threatened" or threatened.

NOTES. *H. macrobotrys* is most likely to be confused with *H. oliveri* (*q.v.*). Some material of *H. pseudofischeri* was determined as this species, possibly on the basis of the subglobose fruit. However, the fruits of *H. macrobotrys* are twice as large as those of the last species, as are the leaves, (3.4–4.6(–7) × 1.5–2.4(–3.2) cm versus (0.9–)2.2–2.8(–4) × (0.5–)0.7–1.1(–2.8) cm. The extremely large measurements for the leaves of *H. pseudofischeri* just referred to derive from an atypical specimen, *Batty* 589 (**T** 3) which may represent a hybrid between the two.

Polhill & Paulo 1303 (**T** 7) appears to be a deviant specimen of *H. macrobotrys*, it is notable for the large size of its leaves, flowers and fruit.

7. **Hermannia oliveri** *K.Schum.* in E.J. 15: 134 (1892) & in E.M. 5: 52 (1900); T.T.C.L.: 600 (1949); U.K.W.F. ed. 2: 97 (1994). Type: Kenya, Teita District: Taveta, 600 m, *Johnston* s.n. (B, holo., presumed destroyed; K! iso.)

Shrub 0.45–1(–2.5) m tall, stems dark or red brown, white stellate-hairy with hairs of two sizes. Leaf-blade ovate or ovate-rhombic, 3.8–6.2 × 1.6–4.7 cm, apex rounded, base obtuse or rounded, margin dentate, 12–25 teeth/side, teeth 0.5–1 × 1.5–5 mm, nerves 5–8 on each side of the midrib, surface with 5–6-armed white-stellate hairs, obscuring ± half the upper surface, lower surface totally obscured; petiole (7–)10–22 mm long; stipules narrowly lanceolate or triangular, 5–7(–14) × 0.5(–3) mm, apex acute. Inflorescence a terminal thyrse 8–13(–22) cm long, densely white stellate-hairy; partial-inflorescences (5–)8–11, (2–)3–7 cm long at base, each bearing (15–)20–30(–65) flowers, lowest partial peduncle 1–3(–18) mm long; peduncular bracts narrowly triangular, 2–3(–7) mm long; partial-rhachis bracts 1–4 mm long; pedicellar bracts filiform, subpatent, single, rarely 2 or 3, 0.5–1.2(–3) mm long; pedicel 2–3.5 mm long. Flowers erect, yellow or orange, 6–8 mm wide. Calyx with basal part campanulate, 2.5–5 × 4–5 mm, lobes spreading slightly, narrowly triangular, 2.7–4 × 1.5–2 mm, stellate-hairy. Petals spatulate, 5–6.5 × 1.8–2.2 mm, not, or slightly involute in basal $^1/_4$, glabrous. Stamens with filaments triangular or elliptic, 1.5 × 0.6–0.8 mm, lacking appendages, glabrous, anthers 4–4.5 × 0.6–1 mm. Ovary ellipsoid, 2.5–3 × 2–2.5 mm, lacking awns; style 3.2–3.5 mm long. Immature fruit capsules ovoid-ellipsoid, 9 × 7 mm, 5-ridged, ridges slightly echinate; seeds unknown.

KENYA. Machakos District: Kiboko Range Research Station, 11 Sep. 1981, *Ndegwa* 96!; Kitui District: Nairobi–Garissa 200 km, Jan. 1961, *Lucas & Williams* 68!; Teita District: Tsavo National Park, Irima Hill, 1 June 1967, *Huck & Huck* 1031!

TANZANIA. Moshi District: Moshi, May 1927, *Haarer* 406!; Pare District: NW spur of N Pare Mts, above Kifaru Estate, 19 May 1968, *Bigger* 1850!; Pare/Lushuto District: Mkomazi Game Reserve, 5 May 1962, *Ibrahim* 652!

DISTR. **K** 4, 6, 7; **T** 2, 3; not known elsewhere

HAB. Grassland, wooded grassland, *Acacia* bushland or woodland; 400–1500 m

SYN. *Mahernia exappendiculata* (Mast.) K.Schum. var. *tomentosa* Oliv. in Trans. Linn. Soc., ser. 2, Bot. 2: 329 (1887), *nom. nud.*

LOCAL USES. None recorded.

CONSERVATION NOTES. Although endemic to the FTEA area, this species is widespread and common (38 specimens are known), and its habitat not significantly threatened. It is rated here as of "least concern" for conservation.

NOTES. The record of this species for Zimbabwe (Wild in F.Z. 1: 548 1961), is erroneous. The specimen concerned, *Phipps* 1330, is *H. macrobotrys*.

The flowers do not open before noon *fide Huck & Huck* 1031 (**K** 7).

8. **Hermannia boranensis** *K.Schum.* in E.J. 33: 311 (1903); Vollesen in Fl. Eth. 2: 180 (1995); Thulin in Fl. Som. 2: 34 (2000). Type: Kenya-Ethiopian border, Boran, Andada, *Ellenbeck* 2153 (B, presumed destroyed, holo.)

Shrub 0.3–0.8(–1.5) m tall, stem 2–4 mm diameter at base, red-brown, subglabrous, with very thinly scattered white 8-armed stellate hairs 0.2–0.6 mm diameter, becoming glabrous, grey, glaucous. Leaf-blade ovate, 3.5–5 × 1.2–4.4 cm, apex rounded, margin crenate to subdentate, base truncate or obtuse, teeth 20–30, 0.5–1 × 1.5–4 mm, nerves 4–7 on each side of the midrib, upper and lower surface moderately densely covered with stellate hairs; petiole 1–3.2 cm long; stipules often early caducous, subulate, 2–4 × 0.6–0.8(–1.5) mm long. Inflorescence a terminal thyrse (6–)18–30 cm long, densely glandular, with scattered simple hairs, partial-inflorescences (8–)12–14(–25), 1.7–3.2(–6) cm long at base, each bearing (3–)5–6(–8) flowers, lowest partial peduncle (10–)12–18(–28) mm long; peduncular bracts filiform, 3–6 mm long; partial-rhachis bracts 2–3 mm long; pedicellar bracts absent; pedicel 2–5 mm long. Flowers suberect, bright yellow, 4–12 mm wide. Calyx membranous, sparsely stellate-hairy, ovoid in bud ± 4 × 3 mm, at anthesis tearing longitudinally into two unequal pieces, revolute, rarely dividing into 5 equal lobes. Petals strongly reflexed, weakly clawed, (6.5–)8–9 × 3–4 mm, blade obovate, claw ± 2 × 1 mm, slightly cupped at base, glabrous. Stamens with filaments oblong, 0.6–1 × 0.5 mm, lacking appendages, glabrous; anthers 4–5 × 0.8 mm. Ovary ellipsoid, 2–2.3 × 1.5–2 mm, lacking awns; style 3–4 mm long. Fruit capsule yellowish brown, 5-angular, ellipsoid in side view, angles with a low ridge, valves 10–12 × 4–4.5 mm, stellately hairy, persistent; seeds subreniform, dull red, 1.5 × 1 × 1 mm, foveolate with scattered papillae.

UGANDA. Karamoja District: Namonono River, Karasuk, Aug. 1962, *J. Wilson* 1271!
KENYA. Northern Frontier District: Mt Kulal, Belessa Kulal, 21 Dec. 1975, *Lamprey et al.* 10! & Wajir, Jan. 1955, *Hemming* 438!; Tana River District: Kora Reserve, Kampi ya Chui–Yumbandei Hill, 27 km, 16 May 1983, *Mungai* 202!
DISTR. **U** 1; **K** 1, 2, 4/7, 7; Ethiopia and Somalia
HAB. *Acacia-Commiphora* bushland; 90–650 m

LOCAL USES. None are recorded.
CONSERVATION NOTES. This species is fairly widespread and common, and its habitat not significantly threatened. It is rated here as of "least concern" for conservation. 15 of the 20 specimens seen for FTEA derive from **K** 1, which appears to be its stronghold.
NOTES. *Hermannia boranensis* is spectacular in flower on account of the large, shuttle-cock-shaped flowers borne on tall, diffuse inflorescences. It is distinguished from other northern Ugandan and Kenyan *Hermannia* species on account of the densely glandular inflorescence and the unusual membranous calyx which usually splits into two and rolls up at anthesis. This is best observed by removing the petals. It is probably most closely related to the geographically distant *H. fischeri* and *H. pseudofischeri* (q.v.).

9. **Hermannia waltherioides** *K.Schum.* in E.J. 33: 312 (1902). Type: Kenya, Machakos [Matichak], *Scott Elliot* 6836 (B? holo., presumed destroyed; K!, iso.)

Shrub 0.3–0.6 m tall; stems purplish black, 2.5–6 mm diameter at base, softly hairy, with stellate hairs, at length glabrous. Leaf-blade greyish green, oblong-elliptic to ovate-rhombic, (2–)2.6–4.2 × (0.6–)0.8–1.5 cm, apex rounded, margin shallowly and obliquely serrate, base cuneate, rarely rounded, nerves 6–8(–10) on each side of the midrib, upper surface densely silvery stellate-hairy, less dense on the lower surface; petiole 3–5 mm long; stipules narrowly lanceolate-subulate, 3–5(–7) mm long. Inflorescence a terminal thyrse (2.5–)3.5–6(–12) cm long, moderately stellate-hairy, partial-inflorescences 4–7, 1–2(–4.5) cm long at base, each bearing (1–)2–5(–7) flowers, lowest partial peduncle 3–7(–20) mm long; peduncular bracts subulate, 4–10 × 1.5–2 mm; partial-rhachis bracts 1–2 mm long; pedicellar bracts absent; pedicel 2.5–6.5 mm long. Flowers erect, yellow, tinged orange, 6–12 mm wide. Calyx base campanulate, 2.5–3 × 3–4.5 mm, lobes narrowly triangular, 3–4 × 1.2–1.8 mm, indumentum as leaf-blade. Petals reflexed, weakly clawed, 5–7 × 3.2–4.5 mm, blade obovate or transversely elliptic, claw 2–3 × 1.8 mm, margins not involute, glabrous. Stamens with filaments 1.3–2 × 0.3–0.8 mm, oblong, with a pair of patent appendages 0.8–1.5 mm from the base, appendages falcate, 0.5–1 × 0.15 mm, glabrous, anthers 4–5 × 0.7–1.3 mm. Ovary ovoid, 2.5–2.8 × 1.3–1.8 mm; style 4 mm long. Immature fruit capsule narrowly ovoid, 5-angular, 13 × 5 mm, slightly stellate-hairy; seeds unknown.

KENYA. Machakos District: 20 Dec. 1931, *van Someren* 1586! & Kilima Kiu, Dec. 1933, *Jex-Blake* 5896! & Chumbi Hill, 30 June 1963, *Verdcourt* 3673!
TANZANIA. Masai District: Loliondo, 13 Nov. 1953, *Tanner* 1812!
DISTR. **K** 4; **T** 2; not known elsewhere
HAB. Thinly wooded stony hillsides; 1500–2000 m

SYN. *Hermannia sp. A* of Agnew in U.K.W.F., ed. 2: 96, f. 25 (1994)

LOCAL USES. None are recorded.
CONSERVATION NOTES. This species is restricted in its distribution. Of the nine specimens known, eight are from Machakos District of Kenya. The other is from northern Tanzania. The Machakos specimens, insofar as detailed locality data are available, are restricted to a small area in the neighbourhood of the Mua hills at the western end of the district. Natural habitat in this area, inhabited by the Akamba, is becoming heavily goat-grazed and is also steadily being cleared for agriculture and habitation (Beentje, pers. comm. 2003). However, much ranching also occurs here, and this is not likely to threaten the species (Luke pers. comm. 2003). *H. waltherioides* is rated here as vulnerable (VU B2ab(iii)) i.e. vulnerable to extinction on the basis of being known from less than 10 locations, with an area of occupancy of less than 2,000 km², and given a continuing decline in the quality of its habitat. An on-the-ground conservation assessment might change the conservation status of the species since it is quite likely that it is now extinct due to clearance at some of its former sites.
NOTES. *Hermannia waltherioides* is closely related and likely to be confused with *H. athiensis* and *H. uhligii* which are also shrubs with *Waltheria*-like, cuneate-based leaves and with, short, dense, terminal inflorescences bearing bright yellow or orange flowers with reflexed petals. However, *H. waltherioides* can be distinguished from these, and all other East African species of the genus by the pair of transverse appendages which make the staminal filaments cruciform. No other species in our area has filament appendages.
    Flowers recorded as aromatic (*Tanner* 1812).

10. **Hermannia athiensis** *K.Schum.* in N.B.G.B. 2: 303 (1899) & in E.M. 5: 49, t. 4D (1900), as *alhiensis*. Type: Kenya, Masai District: Athi [Alhi] Plateau, Feb., *Pospischil* s.n. (B, holo., presumed destroyed)

Shrub (0.1–)0.3–0.6 m tall, stems often spreading, several from a thick, woody underground rootstock, 2–4.5 mm diameter at ground level, stellate-hairy, soon glabrous, glaucous. Leaf-blade lanceolate or lanceolate-rhombic, less usually elliptic, 1.5–3.5 × (0.5–)0.9–1.8 cm, apex rounded, margin serrate, base cuneate or rounded, nerves 5–7 on each side of the midrib, grey velvety hairy on both surface with dense stellate hairs; petiole (2.5–)5–6(–10) mm long; stipules subulate, 2–6 mm long. Inflorescence with internodes contracted, often resembling an umbel, a terminal thyrse 3–5.5 cm long, densely grey stellate-hairy, partial inflorescences 3–6, 1.8–3.5 cm long at base, each bearing 2–3 flowers, lowest partial peduncle 2–5 mm long; peduncular bracts trifid, 4.5–7 mm long; partial rhachis bracts filiform, 1.5–4 mm long; pedicellar bracts filiform, patent 1–2.5 mm long; pedicel 1–3 mm long. Flowers erect, lemon yellow, 14–17 mm wide. Calyx base campanulate, 2.5–4 × 4–4.5 mm, lobes narrowly triangular, 3.5–6 × 1.5–2.5 mm, sparsely stellate-hairy. Petals clawed, 10.5–12 mm long, blade orbicular to widely elliptic, 7.5–8 × 6–7 mm, claw suboblong, 3 × 1.5–2.5 mm, margins not involute, rarely one margin involute (*Verdcourt* 2313), glabrous. Stamen filaments suboblong, 1.6–2 × 0.6–0.8 mm, lacking appendages, glabrous, anther (5–)6–7 × 0.8–1 mm. Ovary subcylindrical to narrowly ellipsoid, 4 × 1.8 mm, style 5–6 mm long. Fruit capsule narrowly ellipsoid–5-winged, 14 × 7–8 mm, apex retuse, wings erose, ± 1 mm wide, sparsely stellate-hairy; seed dull red, subreniform, 1.5 × 1.1–1.3 × 1 mm, foveolate, lacking papillae.

KENYA Laikipia District: 40 km N of Rumuruti, 5 Nov. 1978, *Hepper & Jaeger* 6624!; Masai District: Athi Plains, Nairobi–Namanga 34 km, 21 Dec. 1963, *Verdcourt* 3844! & Ngong Hills, Nairobi–Magadi 48 km, 19 Sep. 1955, *Greenway* 8848!

DISTR. **K** 3?, 6; not known elsewhere

HAB. *Acacia-Commiphora* bushland; 1150–2000 m

LOCAL USES. None are recorded.

CONSERVATION NOTES. This species, known from nine or ten specimens, is entirely restricted to Masai District of **K** 6, apart from a dubious specimen from **K** 3 (see note below). Slow but steady clearance of its natural habitat for habitation and small-scale agriculture is occurring, particularly towards Nairobi. However, much of its range is used for cattle and game ranching, which is not considered likely to be a threat to *Hermannia* (Luke, pers. comm.) It is rated here as vulnerable to extinction (vulnerable (VU B2ab(iii)) on the basis of being known from less than ten locations, with an area of occupancy of less than 2,000 km², and given a continuing decline in the quality of its habitat.

NOTES. *Hermannia athiensis* is closely related to *H. uhligii* and has been considered as merely a large-flowered form of the latter. They are sympatric in **K** 6. The most convincing character for maintaining these two species as distinct are the fruits. In *H. uhligii* these are twisted, the valves contorted and bullate, whereas in *H. athiensis* the fruits are straight, the valves flat. Generally, *H. athiensis* has the larger flowers, described as lemon yellow, held in a dense umbel-like inflorescence. In *H. uhligii* the flowers tend to be described as bright orange yellow; they are smaller, in more diffuse inflorescences. However, without fruits, confusion between these two species is possible. The single specimen of *H. athiensis* from **K** 3 lacks fruits, and may merely be a larger-than-normal flowered version of *H. uhligii*. Unusual specimens previously referred to either of these species from **K** 1 have recently been distinguished as *H. pseudathiensis* (q.v.).

The spelling *H. athiensis* is used here, since *H. alhiensis*, used in the protologue, is meaningless. In the protologue Schumann clearly bases the name on the locality of the only specimen available to him, that of *Pospischil*, cited as "Alhi-Plateau". In her index of collecting localities for FTEA, Polhill (1988: 6) attributes the name Alhi only to Pospischil and treats it as a misnomer for Athi. Engler pointed out Schumann's error in E.J. 55: 355 (1919), pointing out that the name is falsely composed and clearly refers to the Athi Plains, thus *athiensis* should be used. Unfortunately Engler corrupts Schumann's spelling *H. alhiensis* further, to *H. alkiensis*. The name is further misspelt in the latest Index Kewensis CD-ROM, probably due to a scanning error, as *H. albiensis*.

11. **Hermannia pseudathiensis** *Cheek* in K.B. 61: 217 (2007). Type: Kenya, Northern Frontier District: 80 km SW of Wajir, *Bally & A.R. Smith* B14503 "sheet 2/2"(K!, holo., iso!; EA, iso)

Shrub 0.2–0.3 m tall, with a single main stem producing several branches, 3–6 mm diameter at ground level, stellate-hairy, soon glabrous, glaucous, underground rootstock not recorded. Leaf-blade lanceolate or lanceolate-rhombic, less usually elliptic, 1.7–4(–7) × 0.8–1.3(–4) cm, apex rounded, base cuneate or rounded, margin serrate, each side with 12–17 teeth, 0.5–1 × 1.5–5 mm, nerves 4–5(–6) on each side of the midrib, grey velvety hairy on both surfaces, very densely so below, with dense stellate hairs; petiole 6–14(–40) mm long; stipules subulate, 2–4 mm long. Inflorescence racemose, (4.5–)6–9.5 cm long, yellow, white-hairy with stellate hairs, basal internodes elongated, ± 2 cm, partial-inflorescences ascending, 4–7, 2–3 cm long at base, each bearing 1 flower, lowest partial peduncle 20–30 mm long; lower peduncular bracts trifid, 3.5–8 mm long; partial rhachis bracts and pedicellar bracts absent; pedicel u-shaped, (2–)3–4 mm long. Flowers apparently nodding, pale yellow or white, width unknown. Calyx base campanulate, 2–3 × 3–6 mm, lobes narrowly triangular, 6 × 1.5–2 mm, fairly densely white stellate-hairy. Petals weakly clawed, 11–12 mm long, blade widely elliptic, ± 7–8 × 5 mm, claw suboblong, ± 3 × 2 mm, margins not involute, glabrous. Stamen filaments suboblong, ± 1.5 × 0.7 mm, lacking appendages, glabrous, anther ± 5 × 1 mm. Ovary subcylindrical to narrowly ellipsoid, 3.5 × 2.5 mm, ovules 10–15 per locule, style 4.5 mm long. Fruit capsule not seen, immature fruits 3-seeded. Fig. 19, p. 116.

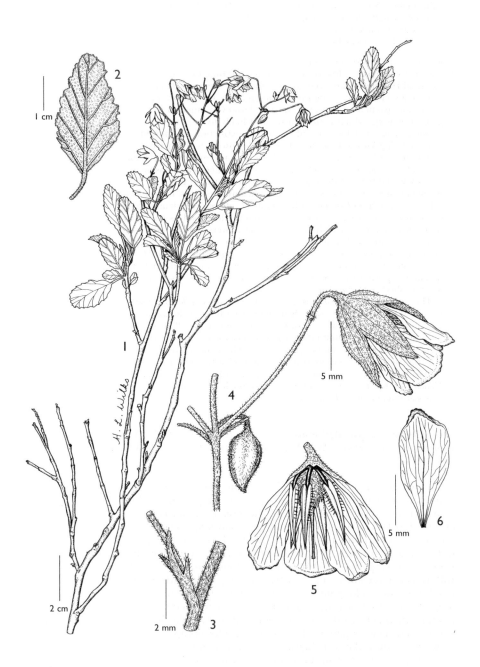

FIG. 19. *HERMANNIA PSEUDATHIENSIS*— **1**, habit; **2**, leaf, lower surface; **3**, apex of peduncle, showing trifid bract; **4**, flower and flower bud, side view; **5**, flower, side view, with petal and two sepals removed; **6**, petal, adaxial view. 1 & 2, 4–6 from *Bally & Radcliffe Smith* B14503, 3 from *Brenan et al.* 14849. Reproduced from Kew Bull. 61: 218. Drawn by Hazel Wilks.

KENYA Northern Frontier District: Wajir–Habaswein road, 29 Nov. 1978, *Brenan et al.* 14849 ! & 26 km NE of Habaswein, 27 April 1978, *Gilbert & Thulin* 1117!; Garissa District: Mado Gashi–Garissa, 13 km S of Mado Gashi, 11 Dec. 1977, *Stannard & Gilbert* 930–960!

DISTR. **K** 1; not known elsewhere

HAB. *Acacia-Commiphora-Cordia* bushland on sandy soils; 200–350 m

LOCAL USES. None are recorded.

CONSERVATION NOTES. This species appears restricted in distribution to the area near Wajir in **K** 1. Overgrazing by e.g. goats is a major problem in this area (Luke, pers. comm. 2003) and may be a threat to this species, though the woody rootstock characteristic of perennial *Hermannia* could afford some protection. Conversely, **K** 1 is so poorly collected that the range of this species may yet prove to be larger than that already known (Luke pers. comm. 2003). *H. pseudathiensis* is rated (Cheek 2004) as VU D2, that is vulnerable to extinction by stochastic changes being known from less than 5 sites (only four specimens are known) with an area of occupancy of less than 20,000 km².

NOTES. *H. pseudathiensis* is known from only the few, poor specimens cited above (of which the best is *Bally & Radcliffe-Smith* B14503). These were previously named as either *H. athiensis* or *H. uhligii*, with which it shares similarities such as trifid basal bracts. *Hermannia pseudathiensis* differs from the last two in being racemose (partial-inflorescences 1-flowered, not several flowered), with a much longer, more diffuse inflorescence (internodes between partial-inflorescences ± 2 cm, not 1 cm or less), in lacking partial-peduncular and pedicellar bracts, in the u-shaped pedicels (not ± straight) and in the generally shorter stipules. More collections of *Hermannia pseudathiensis* are needed to elucidate this species further and to improve its description. Open flowers and mature fruits are entirely unknown in this species.

12. **Hermannia uhligii** *Engl.* in E.J. 55: 354 (1919); T.T.C.L.: 599 (1949); Blundell, Wild Fl. E.Afr.: t. 286 (1987); U.K.W.F. ed. 2: 96, t. 25 (1994). Types: Tanzania, Kilimandjaro–Meru, *Uhlig* 354 (B, presumed destroyed, syntype); Mbuga, *Jaeger* 81 (B, presumed destroyed, syntype); *Merker* 662 (B, presumed destroyed, syntype).

Shrub, usually procumbent, 0.3–0.6 m tall, stems several from a thick, woody underground rootstock, 2–4 mm diameter at base, stellate-hairy, reddish brown, becoming glabrous, grey-glaucous. Leaf-blade lanceolate or lanceolate-rhombic, less usually elliptic, 1.8–3.3(–4) × (0.4–)0.7–1.4(–3.7) cm, apex rounded, margin serrate, base cuneate or rounded, nerves 4–7(–10) on each side of the midrib, grey velvety hairy on both surfaces with dense stellate hairs; petiole (1–)5–9(–15) mm long; stipules subulate, (2–)3.7–5.5(–7) mm long. Inflorescence a terminal thyrse 2–6 cm long, densely grey stellate-hairy, partial inflorescences 3–5, 1–2(–5) cm long at base, each bearing 1–3(–6)flowers, lowest partial peduncle 0.5–7(–19) mm long; peduncular bracts trifid, rarely subulate, 2–6 mm long; partial rhachis bracts filiform, ± 2 mm long; pedicellar bracts filiform, patent 1–3.5 mm long; pedicel 1.2–4 mm long. Flowers erect, yellow, 8–12 mm wide. Calyx base campanulate, 1.7–3 × 2.2–4.5 mm, lobes narrowly triangular, 3–7 × 1.4–2.5 mm, sparsely stellate hairy. Petals clawed, 6–8(–9) mm long, blade widely elliptic, 4–6 × 2.7–3.5 mm, claw suboblong, 2–3 × 0.7–1.3 mm, margins involute, glabrous. Stamen filaments suboblong, 1–1.5 × (0.3–)0.7 mm, lacking appendages, glabrous, anther 3.5–5 × 0.7–1 mm. Ovary subcylindrical to narrowly ellipsoid, 2.2–3.2 × 1.1–1.7 mm, style 3–4.5 mm long. Fruit possibly indehiscent, helically twisted, cylindrical–5-winged, 15 × 6 mm, apex retuse, surface inflated around individual seeds, sparsely stellate-hairy. Immature seed dull red, subreniform, foveolate, lacking papillae.

KENYA. Machakos District: Lukenya Hill, 31 Jan 1971, *Mwangangi* 1580!; Masai District: Siyabei Gorge, 15 July 1962, *Glover & Samuel* 3189!; Tana River District: Baomo village, 13 March 1990, *Luke et al.* in TPR 218 !

TANZANIA. Musoma District: Serengeti, E of Seronera airstrip, 16 May 1962, *Greenway & Myles Turner* 10640!; Mbulu District: Lake Manyara National Park, 8 June 1965, *Greenway & Kanuri* 11832!; Pare/Lushoto Districts: Mkomazi Valley, 30 Nov. 1935, *Burtt* 5319!

DISTR. **K** 3, 4, 6, 7; **T** 1–3; not known elsewhere

HAB. *Acacia-Commiphora* bushland or grassland, sometimes on limestone; 30–1850 m

LOCAL USES. "Roots put in soup" (*Greenway* 4267). Several specimens refer to the species being grazed by animals, e.g."goat-cropped", *Tanner* 4434.

CONSERVATION NOTES. This species is fairly widespread and common (48 specimens seen) in Kenya and Tanzania, and its habitat not significantly threatened over the whole of its range. Several specimens refer to it as being "common" or "locally common", or even (*Semsei* 2135) a "very common shrub". It is rated here as of "least concern" for conservation.

NOTES. *Hermannia uhligii* has twisted and inflated fruits which are otherwise unknown in East Africa. In flower it can be confused easily with *H. athiensis* (q.v.) of which it resembles a smaller form.

13. **Hermannia fischeri** *K.Schum.* in E.J. 15: 134 (1892) & in P.O.A. C: 270 (1905) & in E.M. 5: 49 (1900); T.T.C.L.: 600 (1949). Types: Tanzania (assumed): "Massai-Steppe", *Fischer* 58, 68 (B, syn., presumed destroyed; lecto. selected here, *Fischer s.n.* "Massai land", K! see note below)

Shrub 0.2–0.5(–0.6) m tall, sometimes pyrophytic (*Pocs* 89232/A), stems purplish brown, branching, 2–3(–4) mm diameter at ground level, stellate-hairy. Leaf-blade lanceolate or ovate, 1.5–3.8 × 0.6–2.8 cm, apex rounded, truncate or rounded, margin serrate, base slightly cordate, nerves 4–6 on each side of the midrib, lower surface densely stellate-hairy, upper more sparsely hairy; petiole 0.3–1.1(–1.5) cm long; stipules subulate, 0.5–2 mm long. Inflorescence a terminal thyrse 5.5–9.5(–17) cm long, indumentum a mixture of stellate and glandular simple hairs, partial-inflorescences 12–20(–26), 0.6–1 cm long at base, each bearing 1(–3) flowers (upper partial-inflorescences usually always 1-flowered), lowest partial peduncle 0.8–5.5 mm long; peduncular bracts subulate, (2–)2.5–3.5 mm long; partial-rhachis bracts mostly absent, if present ± 0.2 mm long; pedicellar bracts absent; pedicel 1.5–2.5 mm long. Flowers pendulous, yellow, 7–11 mm wide. Calyx pre-anthesis ovoid, 5–6 × 2–3 mm, sparsely stellate-hairy, lobed for ³/₅, lobes united, at anthesis reflexed, turned inside-out, the basal part then resembling a glabrous convex disc, the lobes mostly becoming free from each other, their inner surface densely white hairy near margin. Petals strongly reflexed, clawed, 5–7 mm long, blade oblong to elliptic, 3–5.5 × 3.5–4 mm, claw suboblong, 1.5–2 × 0.8–1.3 mm base not involute, glabrous. Stamen filament oblong, 0.5–0.6 mm long, lacking appendages, glabrous, anthers 4–4.5 × 1 mm. Ovary oblong-ellipsoid, 2–3 mm long; style ± 4 mm long. Fruit a patent to pendulous (but apex always pointing away from infructescence axis) cylindrical-ellipsoid capsule ± 10 × 5–6 mm, base and apex truncate, stellate-hairy; seeds 9–13 per locule, mature seeds not seen.

KENYA. Kilifi District: Galana Ranch at Dakabuku Hill, 6 April 1975, *Bally* 16659!; Kwale District: Mackinnon Road, 1 Sep. 1953, *Drummond & Hemsley* 4098! & idem, Feb. 1963, *Tweedie* 2570!.

TANZANIA. Masai District: Kibaya steppe, 25 Aug. 1932, unknown Germanic collector 1605!; Moshi District: Kilimanjaro, pre-1884, *Johnston* s.n.!; Lushoto District: Mkomazi Game Reserve, Kamakota Hill, 11 June 1996, *Abdallah et al.* 96/164!

DISTR. **K** 7; **T** 2, 3; not known elsewhere

HAB. *Acacia-Commiphora* bushland on grey sandy soils; 300–550 m

LOCAL USES. None are recorded.

CONSERVATION NOTES. This species is fairly widespread over a large area, though not very common (ten specimens are known), and its habitat not significantly threatened if the whole of its range is taken into account. Clearance of trees for making charcoal has been occurring within the range of this species, but is unlikely to affect *Hermannia* deleteriously. *H. fischeri* is rated here as "near threatened" for conservation.

NOTES. *Hermannia fischeri* is most likely to be confused with *H. pseudofischeri*, q.v. for distinguishing notes and affinities.

The lectotype indicated above is suspected of being a fragment of one of the two syntypes sent by Schumann to Kew just before publication of his protologue. The label is printed "Ex Museo botanico Berolinensis" and is written "*Hermannia fischeri* K.Schum., Massailand, Leg. *Fischer*" all in what resembles Schumann's script. It is stamped "Rec. Dec. 1891".

14. **Hermannia pseudofischeri** *Cheek* in K.B. 61: 215 (2007). Type: Kenya, Kitui/Kilifi District: Lali Hills, Galama River, *Njoroge Thairu* 98 (K!, holo.; EA, iso.)

Shrub 0.3–0.5 m tall, stems red or brown, branching, 2–3 mm diameter at ground level, stellate-hairy. Leaf-blade lanceolate-oblong, (0.9–)2.2–2.8(–4) × (0.5–)0.7–1.1(–2.8) cm, apex rounded, margin serrate, base cordate, truncate or obtuse, nerves 4–7 on each side of the midrib, upper and lower surface densely stellate-hairy; petiole 0.6–1.1(–1.5) cm long; stipules subulate, 1–1.5 mm long. Inflorescence a terminal thyrse 13–16(–26) cm long, indumentum a mixture of stellate and glandular simple hairs, partial-inflorescences 20–44, 0.6–1 cm long at base, each bearing 1–3 flowers (upper partial-inflorescences usually 1-flowered), lowest partial peduncle 1.5–3 mm long; peduncular bracts subulate, 1.5–2 mm long; partial-rhachis bracts usually absent, when present ± 0.2 mm; pedicellar bracts absent; pedicel 2–3.5 mm long. Flowers pendulous, yellow, 5–7 mm wide. Calyx pre-anthesis ovoid, ± 5 × 2.6 mm, sparsely stellate-hairy, lobed for $^3/_5$, lobes united; at anthesis reflexed, turned inside-out, the basal part resembling a convex disc, the lobes mostly becoming free from each other, their inner surface densely white hairy near margin; petals strongly reflexed, clawed, 5–6 mm long, blade obovate to elliptic, 3–4 × 2–3 mm, claw suboblong, 2 × 0.8–1.3 mm, base not involute, glabrous. Stamen filament flattened, oblong in frontal view, 0.5–0.6 mm long, lacking appendages, glabrous, anthers 4–4.5 × 0.5 mm. Ovary globose, 1.8–2 × 1.6 mm; style 3.2–4.2 mm. Fruit a pendulous subglobose capsule 4–5 × 5–6 mm, apex depressed, sparsely stellate-hairy; immature seeds dull red, subreniform, 0.8 × 0.5 × 0.5 mm, foveolate, lacking papillae. Fig. 20, p. 120.

KENYA. Kitui/Kilifi District: Lali Hills, Galana River, 27 March 1963, *Njoroge Thairu* 98!; Lali Hills, Kosai, 22 May 1963, *Parker* Gm/523/H!; Kilifi District: Crocodile camp–Malindi 30 km, 6 June 1986, *Linder* 3658!
TANZANIA. Lushoto District: Mkomazi, April 1937, *Moreau & Moreau* 8 ! & near Same, 1 Sep. 1969, *Batty* 589
DISTR. **K** 4/7, 7; **T** 3; not known elsewhere.
HAB. Dry bushland; 150–900 m

LOCAL USES. None are known.
CONSERVATION NOTES. *Hermannia pseudofischeri* is rated (Cheek 2004) as VU D2, that is vulnerable to extinction by stochastic changes being known from 5 or less sites (only five specimens are known) with an area of occupancy of less than 20,000 km². Within its range it is fairly scarce, given that only five specimens are known and that the only collector who comments on its frequency reports it as "occasional" (*Linder* 3658). Luke (pers. comm. 2003) reports that the Kenyan material has mostly been collected from near the border of Tsavo East N. P., and the Tanzanian material from Mkomazi National Park, and that as both these areas are well protected, ongoing habitat destruction is not likely to be a concern for this species.
NOTES. *Hermannia pseudofischeri* is closely related and most likely to be confused with *H. fischeri*. They are sympatric and have similar ecological requirements. The two are easily distinguished in fruit (subglobose, 4–5 × 5–6 mm, pointing at axis in *H. pseudofischeri*, but ellipsoid-cylindric, ± 10 × 6 mm, pointing away from axis in *H. fischeri*). In flower *H. pseudofischeri* can best be separated as follows: by the inflorescence length 13–16(–26) cm long, not 5.5–9.5(–17) cm; bract length 1.5–2 mm long, not (2–)2.5–3.5 mm long. Both species are distinguished from most other shrubby, yellow-flowered *Hermannia* by their erect, slender, glandular and stellate-hairy inflorescences which bear numerous, usually one-flowered, partial-inflorescences. *Hermannia uhligii* and *H. athiensis* are examples of another pair of species in E Africa separated mainly by their fruits rather than their flowers. The flowers of *H. pseudofischeri* and *H. fischeri*, with their strongly and completely reflexed petals and sepals, resemble a cone. This character and the presence of a glandular inflorescence links the species with *H. boranensis*, and, to some extent *H. macrobotrys*. The last two species do not have 1-flowered partial peduncles, however.

15. **Hermannia paniculata** *Franchet* in Revoil, Fl. Çomalis: 19 (1882); K. Schum. in Engl. Hochgeb. Fl.: 305 (1892) & in E.M. 5: 50 (1900); Vollesen in Fl. Eth. 2: 180, f. 80.7 1–3 (1995); Thulin in Fl. Som. 2: 34 f. 18 A–C (2000). Type: N Somalia, unspecified, *Revoil* 23 (P, holo., *fide* Thulin *loc.cit.*)

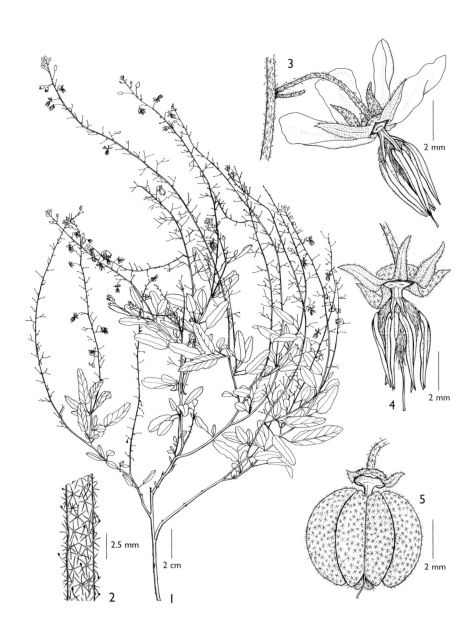

FIG. 20. *HERMANNIA PSEUDOFISCHERI* — **1**, habit; **2**, indumentum of inflorescence rachis; **3**, flower, side view, nearest petal removed; **4**, flower, all petals removed; **5**, fruit, side view. 1–4 from *Linder* 3658; **5** from *Thairu* 98. Reproduced from Kew Bull. 61: 216. Drawn by Hazel Wilks.

Small shrub; stems branching, reddish brown, glaucous greyish purple, ± 3 mm diameter at ground-level, sparsely stellate-hairy, soon glabrous. Leaf-blade ovate or rhombic-ovate, 1.3–3.7 × 0.7–2.4 cm, apex rounded, margin serrate, base cuneate or rounded, nerves ± 6 on each side of the midrib; petiole 3–12 mm long; stipules narrowly ovate, 2–4 × 1.5–2, apex acuminate, base subcordate, sessile, midrib pronounced. Inflorescence a terminal thyrse 5–11 cm long, indumentum a mixture of stellate and glandular simple hairs, partial-inflorescences 10–20, 1–2.5 cm long at base, each bearing 4–8 flowers, lowest partial peduncle 7–9 mm long; peduncular bracts trifid, 3–4.5 mm long; partial-rhachis bracts lanceolate, ± 4 mm long; pedicellar bracts filiform, 0.3 mm long; pedicel 2–3 mm long. Flowers yellow, 4–5 mm wide. Calyx base campanulate, 2 × 3 mm, lobes reflexed at anthesis, triangular, 1.7 × 0.8 mm, densely stellate-hairy. Petals clawed, 3.9–4.3 mm long, blade elliptic, ± 2.5 × 1.5–2.2 mm, claw oblong, ± 1.8 × 0.9 mm, margins involute. Stamen filament subcylindrical, 1.3 × 0.2 mm, lacking appendages, glabrous, anthers 2.2–3.5 × 0.5 mm. Ovary globose to widely ovoid, 1.3 × 1.2 mm, style 4.5 mm. Fruit schizocarpous, ± 2 × 3 mm, mericarps 5, orange-segment-shaped, 2 × 1.3 × 1 mm, ± 3-seeded, stellately hairy; immature seed dull red, reniform, 1.3 × 1 × 1, foveolate, lacking papillae.

KENYA Northern Frontier District: El Wak, 16 Dec. 1971, *Bally & Radcliffe-Smith* in B14641!
DISTR. **K** 1; Djibouti, Ethiopia, Somalia; Arabia
HAB. *Commiphora* bushland; ± 470 m

LOCAL USES. None are recorded.
CONSERVATION NOTES. Although known from only one specimen in the FTEA area, this species is fairly widespread and common and its habitat not significantly threatened over its global range. It is rated here as of "least concern" for conservation.
NOTES. *Hermannia paniculata* is not likely to be confused with other East African shrubby yellow-flowered *Hermannia*, apart from *H. vollesenii* (q.v. for distinguishing characteristics) because these two species have flowers half the size of the smallest of the other species. It is also apparently unique in the area, and possibly in the genus in having a schizocarp with mericarps rather than dehiscent capsules. The fruits of *H. vollesenii* are unknown.

16. **Hermannia volleseni** *Cheek* in K.B. 61: 219 (2007). Type: Ethiopia, Sidamo, Waddera–Negele Rd, Bittata, 3 km N, *Gilbert et al.* 8064 (K!, holo., ETH, iso.)

Prostrate shrub ± 0.4 m tall; stems to at least 30 cm long, purplish brown, stellately hairy. Leaf-blade obovate or ovate, 2.2–4.3 × 1.1–2(–3) cm, apex rounded, margin serrate, base cuneate, rarely rounded or truncate, nerves 6–8 on each side of the midrib, both surfaces densely hairy with stellate hairs; petiole 5–9 mm long; stipules lanceolate, 3.5–7 × 2.5–3.5 mm. Inflorescence a terminal thyrse 10–25 cm long, indumentum a mixture of stellate and glandular simple hairs; partial-inflorescences 7–16, 1.7–4 cm long at base, each bearing 5–8 flowers, lowest partial peduncle 1.5–10 mm long; peduncular bracts lanceolate, 4–9 × 2–2.5 mm; partial-rhachis bracts 1.5 mm long; pedicellar bracts filiform, 1.5 mm long; pedicel 2–2.5 mm long. Flowers pendulous, yellow. Calyx base inflated, campanulate, 1.8 × 4.5 mm, lobes with margins deeply concave, 3 × 3 mm, densely stellately hairy. Petals clawed, 5–6 mm long, blade transversely elliptic, 3–3.5 × 3.5–4 mm, claw oblong, 2–2.25 × 0.8 mm, margins involute on one side, glabrous. Stamen filaments subterete, 2–2.3 × 0.2 mm, lacking appendages, glabrous, anther 2.4–2.8 × 0.7 mm. Ovary widely ovoid, 1.5 × 1.5 mm, style 5 mm long. Fruit and seed unknown.

KENYA Northern Frontier District: Moyale, 24 April 1952, *Gillett* 12924! & Wajir–Tarbaj, 40 km, 19 May 1972, *Gillett* 19734!
DISTR. **K** 1; Ethiopia (Sidamo)
HAB. Open grassland and *Acacia-Commiphora* bushland; 300–1200 m

SYN. *H. sp. A aff. paniculata* Vollesen in Fl. Eth. 2: 180 (1995)

LOCAL USES. None recorded.

CONSERVATION NOTES. This species, known from only seven specimens, is restricted to S Ethiopia (Sidamo Province) and northern Kenya. Overgrazing in the Wajir area of northern Kenya is likely to be a problem: see notes for *H. pseudathiensis* (Luke, pers. comm. 2003). No information is available to me on its status in Ethiopia. It is rated here as vulnerable to extinction (vulnerable (VU B2ab(iii)) on the basis of being known from less than ten locations, with an area of occupancy of less than 2,000 km², and given a continuing decline in the quality of its habitat.

NOTES. Closely similar to the sympatric *H. paniculata* in general appearance and flower size, *H. vollesenii* is distinguished by its prostrate (not erect) habit, its compact, long, slender (6–10 × as long as wide) inflorescence (not diffuse, ± as wide as long), its densely and conspicuously glandular inflorescence indumentum (not sparsely and inconspicuously glandular) and in the ligulate-lanceolate (not ovate) stipules.

INCERTAE SEDIS

Hermannia conradsiana Engl., *H. stuhlmannii* K.Schum., *H. volkensii* K.Schum. and *H. stuhlmannii* Engl. are described from our area but, in the absence of original material (destroyed in Berlin in 1943) or illustrations, their placement is uncertain. No specimens at K have been attributed to these names. The original protologues were imprecise regarding the characters needed to place these names. *H. conradsiana* (based on *Conrads* 32, Mwanza District: Usukuma, Bukumbi, **T** 1) was maintained by Brenan (T.T.C.L.: 600, 1949), but Gillett *in litt.* suggested it may be an insignificant variant of *H. exappendiculata*. *H. volkensii* (*Volkens* 2134, **T** 2) was treated as a synonym of *H. oliveri* by Brenan (*loc. cit.*). *H. stuhlmannii* K.Schum. (*Stuhlmann* 360, **T** 6) was accepted by Brenan in T.T.C.L.: 600 (1949). It may be a synonym of *H. macrobotrys*. *H. stuhlmannii* Engl., a later homonym, based on *Stuhlmann* 285, has been attributed to *H. tigreensis* by Gillett *in litt.* There is no evidence that either Gillett or Brenan saw the specimens concerned and in the absence of original material being discovered these names are best ignored.

15. **MELOCHIA**

L., Sp. Pl.: 674 (1753) & Gen. Pl., ed. 5: 304 (1754); Goldberg, Contr. U.S. Nat. Herb. 34(5): 191–363 (1967)

Herbs, subshrubs, shrubs or small trees (outside FTEA), pubescent with stellate, simple, bifurcate or glandular hairs. Leaves simple, irregularly crenate-serrate, 3–5(–7)-nerved from the base, petiolate, stipulate, stipules caducous or persisting. Inflorescences axillary or terminal cymes, flowers sometimes congested; bracts conspicuous, persistent. Flowers bisexual, distylous or homostylous, pedicellate. Calyx campanulate, shortly 5-lobed, persistent, sometimes accrescent. Petals 5, flattened throughout (not cucullate), without appendages, unguiculate, free or briefly adnate to the base of the staminal tube, marcescent. Stamens 5, antepetalous, filaments united into a staminal tube; anthers 2-celled, cells parallel. Staminodia absent or rarely present (outside FTEA). Ovary 5-celled, sessile or briefly stipitate; ovules (1–)2 per locule, superposed, anatropous, bitegmic; styles 5, free or connate at the base, filiform. Capsule 5-valved, depressed globose or pyramidal and 5-winged, loculicidal or septicidal (or both), often surrounded by the persistent calyx and marcescent corolla; seed small, ascending, obovoid with the dorsal surface rounded and two sides flattened, endospermous; embryo straight, flat, cotyledons suborbicular, subcordate.

A genus of 50–60 species found in tropical and subtropical regions, with centers of diversity in Malesia and the Pacific (sect. *Visenia*), and Central and South America (sects. *Melochia*, *Mougeotia*, *Physodium*, and *Pyramis*), and a number of species now pantropical weeds.

Flowers in terminal (rarely axillary) cymes; stems often with a
 line of stellate hairs decurrent from the stipule bases .... 1. *M. corchorifolia*
Flowers in axillary cymes; stems and young parts densely
 covered with long silky hairs or at most uniformly
 pubescent with rather short hairs .... 2. *M. melissifolia*

1. **Melochia corchorifolia** *L.*, Sp. Pl. 675 (1753); Mast. in F.T.A. 1: 236 (1868); K. Schum. in Engl., P.O.A. C: 271 (1895); Th. Durand & Schinz in Acad. Roy. Belg., Mém. 53(4): 81 (1896); K. Schum. in E.M. 5: 41, t. 3/G (1900); De Wild., Pl. Bequaert. 1: 515 (1922); F.W.T.A., App.: 108 (1937); F.P.S. 2: 6 (1952); F.W.T.A., ed. 2, 1(2): 318 (1958); Wild in F.Z. 1(1): 535, t. 100/B (1961); Germain in F.C.B. 10: 256 (1963); Goldberg in Contr. U.S. Nat. Herb. 34(5): 304 (1967); U.K.W.F.: 193 (1974); Wild & Gonç. in Fl. Moçamb. 27: 23, t. 3/B (1979); Vollesen in Opera Bot. 59: 34 (1980); Vollesen in Fl. Eth. 2(2): 178, fig. 80.6/1–5 (1995); Thulin, Fl. Som. 2: 30, fig. 16/A-E (1999). Lectotype ("as type"), selected by Goldberg (1967: 305): India, "Melochia corchori folio," Dillenius, Hort Eltham.: t. 176, fig. 217 (1732), based on a specimen in Herb. Dillenius (OXF)

Annual or perennial herb or subshrub, 0.4–2.5 m tall; stems erect or decumbent, with scattered stellate and sometimes also bifurcate and simple hairs, often with a line of stellate hairs decurrent from the stipule bases. Leaves narrowly ovate or ovate to oblong-lanceolate, 1.5–7.5 cm long, 0.8–3.8 cm wide, apex acute, margin irregularly crenate-serrate, base cuneate to truncate, membranaceous, sparsely hispidulous, a few simple (rarely stellate) hairs on the midrib above and veins below; 5-nerved; petiole 0.4–3.8 cm long, pubescent on the upper surface at least with mostly simple but also bifurcate and stellate hairs; stipules subulate-lanceolate, 6–10 mm long, 1–2 mm wide, margins ciliolate, often persisting. Inflorescences dense terminal or rarely axillary cymes with 5–25 flowers each; peduncle to 7.5 cm long, hispidulous; bracts numerous, 8–9 mm long, 0.5–2 mm wide, outer bracts stipule-like, subulate-lanceolate to falcate, inner ones linear to filiform, often the 4 bracts immediately subtending the calyx forming an involucre, margins ciliolate. Flowers homomorphic (all brevistylous in material examined). Calyx campanulate, 2–3.5 mm long, hispidulous without, glabrous within, the lobes widely deltoid, 0.4–0.6 mm long, 0.8–1 mm wide, with abruptly acuminate teeth, margins ciliolate. Petals obovate to oblanceolate, shortly clawed, 3–5.5(–8) mm long, 1–2.5 mm wide, white with yellow (or green) throat, rarely yellow, pink or pale pink (Zanzibar material), glabrous. Stamens adherent to the petals below, or almost entirely free, ± 2–2.5 mm long, filaments united in a tube almost to apex, ± 2 mm long, free portion of filaments at most 1 mm long. Ovary oblong-ovoid, 1.5–2(–3) mm long, 1–2 mm in diameter; densely pilose; styles ± 1–1.5 mm long, free or sometimes slightly connate below, papillate. Capsule depressed-globose with 5 ± rounded angles, 4–5 mm long, 4–5 mm in diameter, hispidulous; 1(–2) seed per loculus; seed ± trigonal with rounded dorsal and two flat surfaces, 2–2.5 mm long, 1–1.5 mm wide, blackish-brown or greyish, testa minutely striate. Fig. 21, p. 124.

UGANDA. Bunyoro District: Bubwe, June 1943, *Purseglove* 1595!; Busoga District: Lake Victoria, Lolui Island, 11 May 1965, *Jackson* 411565!; Mengo District: Kitabe near Entebbe, July 1935, *Chandler* 1262!
KENYA. Kilifi District: Kibarani, 9 Apr. 1946, *Jeffrey* K.518!; Kwale District: Kivumoni, near Shimoni, 20 Aug. 1953, *Drummond & Hemsley* 3928!; Tana River District: Tana River National Primate Reserve, Baomo Lodge Forest, 17 Mar. 1990, *Luke et al.* T.P.R. 522!
TANZANIA. Tanga District: Amboni, June 1893, *Holst* 2906!; Uzaramo District: Kunduchi, 19 May 1968, *B.J. Harris* 1778!; Songea District: Mbambe Bay, 5 Apr. 1956, *Milne-Redhead & Taylor* 9533!; Zanzibar I., Kisimbani, 17 July 1962, *Faulkner* 3075!
DISTR. **U** 2–4; **K** 7; **T** 1, 3–6, 8; **Z**; **P**; tropical and subtropical Africa, Asia, and Australia; introduced in tropical and subtropical America
HAB. A weed of cultivation, roadsides, and grassland, often in wet areas; 0–950 m

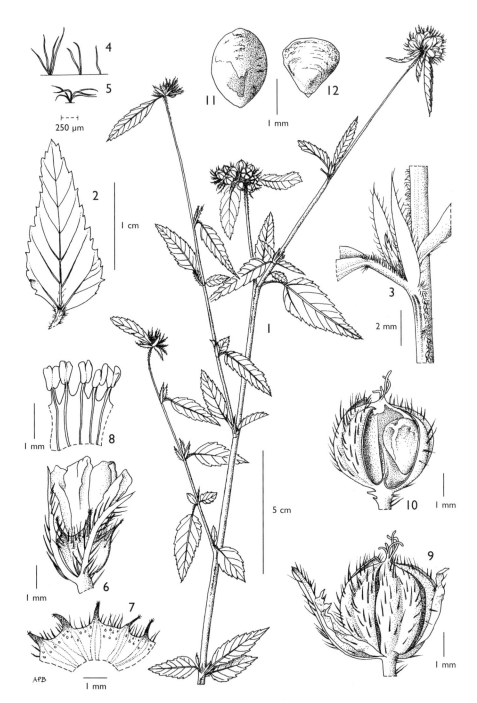

FIG. 21. *MELOCHIA CORCHORIFOLIA* — **1**, habit; **2**, adaxial surface of leaf; **3**, stem node showing stipules and stem hairs; **4**, hairs from leaf petiole; **5**, stellate hairs from stem; **6**, flower, lateral view, with epicalyx; **7**, calyx opened out, showing inner surface; **8**, staminal tube opened out, showing inner surface; **9**, fruit, dehisced, with persistent petals; **10**, fruit, one valve removed to show seed; **11**, seed, ventral surface; **12**, seed, end view. All from *Drummond & Hemsley* 3928. Drawn by Andrew Brown.

SYN. [*M. melissifolia* sensu De Wild., Pl. Bequaert. 1: 515 (1922) *pro parte*]

LOCAL USES. In Kenya, the leaves are eaten as a vegetable (*Graham* 1717). They are also used to treat unspecified stomach diseases (*Jeffrey* K.518).

CONSERVATION NOTES. A widespread and common weedy species, here assessed as of Least Concern for conservation.

2. **Melochia melissifolia** *Benth.* in Hook., J.B. 4: 124 (1841), as *melissaefolia*; Mast. in F.T.A. 1: 236 (1868) pro parte? [Senegambia, *Heudelot* & Niger, *Barter* cited]; K. Schum. in Engl., P.O.A. C: 271 (1895); De Wild., Pl. Bequaert. 1: 515 (1922); F.W.T.A., ed. 2, 1(2): 318 (1958); Wild in F.Z. 1(1): 535 (1961); Germain in F.C.B. 10: 257 (1963); Goldberg in Contr. U.S. Nat. Herb. 34(5): 293 (1967). Types: Guyana, "British Guiana, rocky situations", *Schomburgk* 366 (K!, syn.), and French Guiana, *Leprieur* 122 (K!, US, syn.)

Annual or perennial herb or subshrub, 0.5–1.5 m tall; stems erect or decumbent, pubescent or hispidulous, densely covered with simple hairs to ± 3 mm long, sometimes also with stellate, bifurcate or glandular hairs. Leaves lanceolate to ovate or rarely almost hastate, often larger and smaller leaves present at a node, the larger 3.5–7.2 cm long, 1.3–4.8 mm wide, the smaller 1.6–3 cm long, 0.8–2 cm wide, apex acute, margin irregularly serrate-crenate, base subcordate or truncate, membranaceous, appressed-hispidulous above and below; 5(–7)-nerved below; petiole 0.4–1.2(–2) cm long, ± pulvinate apically, pubescent with some gland-tipped hairs or hispidulous; stipules subulate-lanceolate, 8–12 mm long, 1–3 mm wide, hirsute, margins ciliate, often persisting. Inflorescences dense axillary cymes; peduncle to ± 5 mm long; bracts numerous, 6.5–9 mm long, ± 0.8 mm wide, outer bracts stipule-like, inner ones linear to filiform, the 3 bracts subtending the calyx forming an involucre, margins ciliolate. Flowers homomorphic (all longistylous in material examined). Calyx widely campanulate, 2.5–4 mm long, sparsely hispid without, glabrous within, the lobes deltoid, 0.6–1.2 mm long, 1.5–2.2 mm wide, with abruptly acuminate teeth. Petals obovate to narrowly obovate, shortly clawed, 9–10.5 mm long, 3–4 mm wide, white (or cream) with a yellow throat, glabrous. Stamens adherent to the petals below, filaments united into a tube 3.7–4.5 mm long, free portion of filaments negligible. Ovary narrowly ovoid, 1.5–2.5 mm long, 0.5–1 mm in diameter, sericeous; styles 3–5.25 mm long, free. Capsule depressed-globose, 3–3.5 mm long, 2–3 mm in diameter, sparsely pubescent, ± hispid apically; 1 seed per loculus; seed ± trigonal with a rounded dorsal and two flat surfaces, 1.5–1.8 mm long, ± 1 mm wide, grey, testa minutely punctate.

DISTR. South America (the Guianas and neighbouring Brazil) and introduced into W Africa

var. **mollis** *K.Schum.* in E.M. 5: 43 (1900); F.W.T.A., ed. 2, 1(2): 318 (1958); F.P.U.: 58 (1971); Vollesen in Opera Bot. 59: 34 (1980). Types: Cameroon, Rio del Rey, comm. June 1887, *Johnston* s.n. (B†, K!, syn.)*; Congo-Kinshasa, upper Congo R., Lulua R., *Pogge* 12 (B†, syn.); Sudan, Ngama on Roab R., *Schweinfurth* 2771 (B†, K!, syn.), Boddo R., Niam-niam country, *Schweinfurth* 3036 (B†, syn.); Uganda, Masaka District: Kalungu, *Scott Elliot* 7359 (B†, K!, syn.)

Petiole ¹⁄₄ the length of the leaf blade, hispidulous.

UGANDA. Mengo District: Kanyanya Valley, 20 Jan. 1983, *Rwaburindore* 1146! & Namanve Swamp, Kiagwe, Mar. 1932, *Eggeling* 575! & Kivuvu, May 1914, *Dümmer* 881!

* Schumann (1900: 43) cited this specimen under both *Melochia melissifolia* var. *mollis* and var. *bracteosa*.

KENYA. Kwale District: Gongoni Forest Reserve NE side, 2 June 1990, *Robertson & Luke* 6345! & Kwale, *Graham* 1900! & Buda Mafisini Forest Reserve, 5 Oct. 1999, *Luke & Luke* 5995
TANZANIA. Mwanza District: Buchosa-Buganda Chiefdoms, NW Uzini, 17 June 1937, *B.D. Burtt* 6576!; Rufiji District: Mafia Island, Kipora-Baleni, 31 Aug. 1937, *Greenway* 5198!; Njombe District: Ruhudje area, Massagati, June 1931, *Schlieben* 1092!; Zanzibar: Kisambani, 17 July 1962, *Faulkner* 3071!
DISTR. U 1, 4; **K** 7; **T** 1, 4, 6, 7; **Z**; tropical Africa, also Madagascar and ?the Mascarene Islands
HAB. A weed of swamps and swamp edges, and less frequently sandy soils; 30–1250 m

SYN. *M. bracteosa* F.Hoffm. in Beitr. Kenntn. Fl. Centr. Ost -Afr.: 13 (1889); Th. & H. Durand, Syll.: 65 (1909); Goldberg in Contr. U.S. Nat. Herb. 34(5): 291 (1967). Type: Tanzania, Mpanda District: Ugalla R., *Hoffman* 275 (B†, holo.)
   M. melissifolia Benth. var. *brachyphylla* K.Schum. in E.M. 5: 43 (1900). Types: Sierra Leone, Falaba, *Scott Elliot* 5095 (B†, syn.; K!, isosyn.); Angola, Malandsche, *Mechow* 491 (B†, syn.; BD, isosyn.); Tanzania, Uzaramo area, *Stuhlmann* 7155ᵃ (B†, syn.)
   M. melissifolia Benth. var. *bracteosa* (F.Hoffm.) K.Schum. in E.M. 5: 43 (1900); F.W.T.A., App.: 108 (1937); F.W.T.A. ed. 2, 1(2): 318 (1958); Vollesen in Opera Bot. 59: 34 (1980); Cable & Cheek, Pls. Mt Cameroon 134 (1998); Aké Assi in Boissiera 58(2): 150 (2002)
   M. mollis Hutchinson & Dalziel in F.W.T.A. 1(2): 250 (1928), *nomen* & in K.B. 1928: 297 (1928); F.P.N.A. 1: 608 (1948); F.P.S. 2: 6 (1952). Type: Sudan, on the Roah River, *Schweinfurth* 2771 (K!, holo.; B†, iso.); *non* (Kunth) Triana & Planch.
   [*M. melissifolia* sensu Wild in F.Z. 1(2): 535 (1960); Germain in F.C.B. 10: 257 (1963); Wild. & Gonç. in Fl. Moçamb. 27: 23 (1979), *non* Benth.]

CONSERVATION NOTES. A widespread and common weedy species, here assessed as of Least Concern for conservation.
NOTE. Circumscription of *M. melissifolia* is problematic and deserves further study. Schumann (1900) divided it into six varieties; the nominate one from South America and the remaining ones from Africa. Goldberg (1967: 293) also recognized *M. melissifolia*, but stated that it was "apparently indigenous to America and [West] Africa." He placed all of the African varieties described by Schumann in synonymy under *M. bracteosa* F.Hoffm. The principal character, however, that Goldberg used in his key to distinguish the two species is not practical in the herbarium. *Melochia melissifolia*, as opposed to *M. bracteosa*, was stated to have homomorphic (versus heteromorphic) flowers. Inasmuch as heterostyly is a populational phenomenon this cannot be easily divined from ex situ. More recently, *M. melissifolia* has been treated as a single, polymorphic species (e.g., Wild, 1960; Germain, 1963) or Schumann's five African varieties have been reduced to two: *M. melissifolia* var. *mollis* and var. *bracteosa* (e.g., Vollesen, 1980).
   Melochia melissifolia var. *microphylla* cannot be distinguished from the type of *M. melissifolia* and is here considered a synonym of same, and presumably introduced from South America into West Tropical Africa. The remaining African varieties described by Schumann intergrade amongst each other, but nonetheless are very different from the American type. "They have flowers twice as large as *M. melissifolia* and their leaves are larger, often hairier, and usually have relatively shorter petiole" (Goldberg, 1967). It seems reasonable to recognize them at varietal rank, and they are here subsumed under *M. melissifolia* var. *mollis*. Most of the material from East Africa can be referred to this variety. Only a handful of specimens deviate in having almost exclusively relatively short pubescence on the stems, which has been used to define var. *bracteosa*, but careful examination invariably reveals the presence of at least a few ± 3 mm long simple hairs on the stems.

UNCERTAIN TAXON

Melochia ? mollis (Kunth) Triana & Planch.
   A specimen labeled "E. Trop. Africa, Dr. Stewart. Com. Dr. Masters, 11/68" appears to be *Melochia mollis* (Kunth) Triana & Planch., which is native to South America and which has not otherwise been collected in Africa or Madagascar. It is doubtful that the material is from FTEA since the Rev. James Stewart, M.D. (1821–1905) collected plants on the Zambesi from 1862–73.

16. **WALTHERIA**

L., Sp. Pl. 673 (1753) & Gen. Pl., ed. 5: 304 (1754); Bayer & Kubitzki in Fam. Gen. Vasc. Pls. 5: 242 (2003)

Shrubs or subshrubs, rarely herbs, pubescent with stellate and/or simple hairs. Leaves simple, crenate-dentate or serrate, palmately 3–5-nerved from the base; petiolate or subsessile; stipulate, stipules usually caducous. Inflorescences axillary or pseudo-terminal cymose clusters, sometimes heads of flowers racemosely or paniculately disposed; bracts small, deciduous. Flowers bisexual, distylous or homostylous, sessile. Calyx 5-lobed, persistent. Petals 5, flattened throughout (not cucullate), without appendages, unguiculate, adnate to the base of the staminal tube, marcescent. Stamens 5, antipetalous, filaments united at the base into a staminal tube; anthers di-thecate, thecae parallel; staminodia absent. Ovary syncarpous, 1-celled, sessile (stipitate in teratological flowers); ovules 2, superposed, anatropous, bitegmic; style solitary, excentric, stigma clavate or plumose. Capsule 2-valved, 1(–2)-seeded, loculicidal or operculate, surrounded by the persistent calyx and marcescent corolla; seed small, ascending, endospermous; embryo straight, flat, cotyledons reniform, suborbicular or oblong, cordate, usually chlorophyllous.

A genus of 50–60 species with its greatest diversity in tropical America (especially Mexico and Brazil) but also with species indigenous to Africa, Asia, Polynesia, and Australia; one species now a pantropical weed.

Waltheria indica *L.*, Sp. Pl. 673 (1753); R. Br. in Narrat. Exped. Congo, App. [Tuckey] 5: 478, 484 (1818); DC., Prodr. 1: 493 (1824); F.P.S. 2: 8 (1952); F.W.T.A., ed. 2, 1(2): 319 (1958); Wild in F.Z. 1(2): 536–537, t. 100, fig. A (1961); Germain in F.C.B. 10: 258–259 (1963); F.P.U.: 58 (1971); U.K.W.F., ed. 2: 191, 193, fig. p. 192 (1974); Wild & Gonç. in Fl. Moçamb. 27: 25–26, t. 3A (1979); Vollesen in Opera Bot. 59: 34 (1980); Troupin, Fl. Rwanda 2: 403 (1983); Vollesen in Fl. Eth. 2(2): 178, fig. 80.5/6–9 (1995); Friis & Vollesen, Biol. Skr. Dan. Vid. Selsk. 51(1): 146 (1998); Thulin, Fl. Som. 2: 32, fig. 16/F-I (1999). Lectotype (as "type"), selected by I. Verd. (1981: 275): "Habitat in India" (LINN no. 852.2, lecto.)

Shrub, subshrub or suffrutescent herb, 0.5–1.5 m tall; stems ascending to decumbent; branches densely stellate-pubescent to tomentose throughout, becoming ± glabrous. Leaves ovate to ovate-oblong or ovate-lanceolate, 2.2–13(–15) cm long, 1–8.5 cm wide, reduced in size below the inflorescence, apex rounded, obtuse or subacute, margin crenate-serrate, base rounded, truncate or subcordate, chartaceous, 5-nerved from the base; ± velvety stellate-pubescent to tomentose above and below, occasionally the hairs stiff, slightly discolorous and paler below; petiole 0.4–4.5 cm long, 1–1.5 mm wide (leaves subtending the inflorescence subsessile), stellate-pubescent; stipules filiform, 3–8(–10) mm long, caducous. Inflorescences axillary, tightly congested (rarely sublax) cymes, occasionally more elongate and paniculiform, peduncle to 4.5 cm long; bracts linear-lanceolate, 3.5–6 mm long, hirsute outside. Flowers homostylous. Calyx turbinate-campanulate, 2.5–3.5 mm long, 10-nerved, stellate-sericeous outside, glabrous within, lobes triangular-subulate, 0.7–1.3 mm long, 0.5–0.3 mm wide. Petals obovate to spathulate, shortly clawed, 3–4 mm long, 0.5–1 mm wide (at apex), yellow or orange-yellow, turning reddish-brown at maturity, glabrous. Staminal tube 1.2–2 mm long; free portion of filaments ± 0.7 mm long, glabrous; anthers erect (abnormal flowers horizontal). Ovary oblong-ellipsoid, flattened apically, 1.5–2.5 mm long, 1–1.5 mm in diameter (widest apically), sericeous apically; style ± 1.5–2 mm long, sericeous, sometimes contorted; stigma plumose-papillate. Capsules obliquely

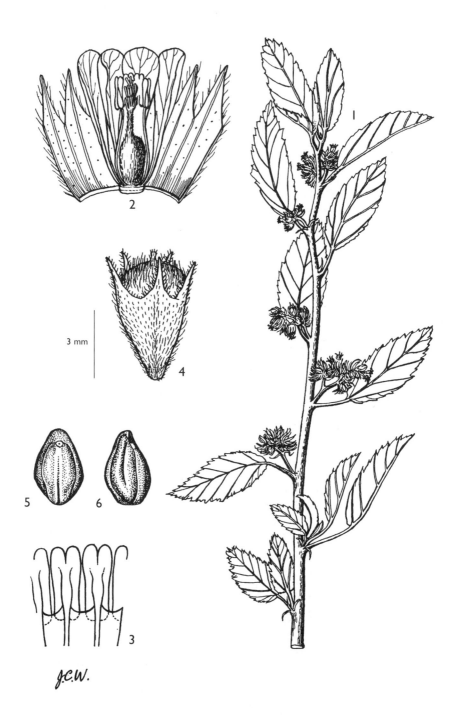

FIG. 22. *WALTHERIA INDICA* — **1**, habit × ²/₃; **2**, flower with calyx opened out and one petal and stamen removed × 6; **3**, anther insertion at top of staminal tube × 12; **4**, fruit, side view × 4; **5**, seed, ventral view × 6; **6**, seed, side view × 6. 1–3 from *Brass* 18010; reproduced from F.Z. 1: t. 100; drawn by Joanna Webb. 4–6 from *J. Wilson* 519, drawn by Hazel Wilks.

obovoid, 2–4 mm long, 2–4 mm in diameter (widest apically), thin-walled, sericeous; 1 (rarely 2)-seeded; seed obovoid, 1.5–2 mm long, 1.5–2 mm in diameter, dark brown or black, testa smooth. Fig. 22, p. 128.

UGANDA. Karamoja District: Chepukui R., Aug. 1962, *J.Wilson* 1289!; Busoga District: Buyindi Hill, 22 May 1951, *G.Wood* 184!; Mengo District: Mpogo, 20 Jan. [without year], *Dummer* 4383!
KENYA. Embu District: Matiri, 19 June 1985, *Beentje* 2166!; Kitui District: Endau Forest, 29 Jan. 1962, *Mbonge* 27!; Kwale District: Lango ya Mwagandi, 9 Feb. 1968, *Magogo & Glover* 31!
TANZANIA. Nusoma District: Kiagata, 15 Mar. 1959, *Tanner* 4008!; Mbulu District: Tarangire, 22 Apr. 1959, *Mahinda* 523!; Tanga District: Tanga–Moa road, 12 Nov. 1964, *Semsei* 3906!; Zanzibar: Mangapwani, 23 Jan. 1929, *Greenway* 1125!
DISTR. **U** 1-4; **K** 1-5, 7; **T** 1–8; **P**; **Z**; a pantropical weed
HAB. A common weed of dry grassland, cultivation, and other disturbed places; 0–1650(–2350) m

SYN. *W. americana* L., Sp. Pl. 673 (1753); DC., Prodr. 1: 492 (1824); Mast. in F.T.A. 1: 235 (1868); K. Schum. in Engl., P.O.A. C.: 271 (1895); Th. Durand & Schinz in Acad. Roy. Belg., Mém. 53(4): 80 (1896); K. Schum. in Ann. Ist. Bot. Roma 7: 36 (1897); K. Schum. in E.M. 5: 45–47, t. 3J (1900) (as "*Typus* K. Schum."); Th. & H. Durand, Syll.: 74 (1909); De Wild., Pl. Bequaert. 1: 516 (1922); Broun & Massey, Fl. Sudan: 123 (1929); Chiov., Fl. Somala: 105 (1929); Cufod. in Miss. Biol. Borana, Racc. Bot. 4: 134 (1939); T.S.K.: 44 (1936); Glover, Prov. Checklist Brit. It. Somaliland Trees, Shrubs Herbs 255 (1947); F.P.N.A. 1: 609 (1948); T.T.C.L.: 604 (1949). Lectotype ("as type"), selected by I. Verd. (1981: 275): "Habitat in Bahama, Barbiches, Surinamo" (LINN, no. 852.1, lecto.)
    W. americana L. var. *densiflora* K.Schum. in E.M. 5: 46 (1900); T.T.C.L.: 604 (1949). Types: Tanzania, Usambara-Usagara: Mtalo, *Holst* 469 (B†, syn.); Uzaramo [Usaramo], *Stuhlmann* 6452 (B†, syn.); Bagamoyo District: near Mtoni, *Stuhlmann* 7184 (B†, syn.); Zanzibar, sine loc., *Peters* 32 (B†, syn.) & idem, *Schmidt* s.n. (B†, syn.), etc. [additional syntypes from elsewhere in Africa and Ascension Island]
    W. americana L. var. *?indica* (L.) K.Schum. L. var. *?* *comb. superfl.*
    W. americana L. var. *?subspicata* K.Schum. L. var. *?* Types: Kenya, Teita District, N'dara, *Hildebrandt* 2468 (B†, syn., K, isosyn.); Tanzania, Tanga, *Holst* 2105 (B†, syn.; K, isosyn.); Uzaramo District: Mguta, *Stuhlmann* 700 (B†, syn.) & Dar es Salaam, *Stuhlmann* 7652 (B†, syn.), 7916 (B†, syn.), etc. [additional syntypes from Sudan, South West Africa, Angola and Mozambique]
    Melochia corchorifolia L. var. *densiflora* K.Schum. in Broun & Massey, Fl. Sudan: 123 (1929), *sphalm. pro W. americana* var. *densiflora* K.Schum.
    Waltheria americana L. var. *sahelica* Roberty in Bull. I.F.A.N. 15: 1403 (1953). Type: Sudan, near El Obèid [Al Ubayyid], *Roberty* 5279 (IFAN, holo.)
    W. laxa Thulin in Nordic J. Bot. 19: 13, fig. 1 (1999); Thulin, Fl. Som. 2: 32, fig. 17 (1999). Type: Somalia, Shabellaha Dhexe Region, Balcad, 25 Nov. 1986, *Raimondo & Fici* A10/86 (PAL, holo.)

LOCAL USES. Numerous medicinal and incidental uses are recorded for this ubiquitous weed. In Kenya, leaves are used as a substitute for toilet paper (*Graham* 1525; *Magogo & Glover* 509) and stems are used as tooth brushes (*Glover et al.* 669). In Tanzania, roots are chewed to treat gonorrhea (*Tanner* 4008) or cough (*Tanner* 4068), and a decoction is drunk to treat painful urination (*Tanner* 3880). Bathing in roots pounded and soaked in cold water is used to treat leprosy, while smoking flour made from the roots is thought to help when the throat has been infected with the same disease (*Tanner* 1239). Leaves soaked in water with cattle urine are used for erysipelas (*Tanner* 583) or a cold infusion of leaves is used to bathe infants to reduce fever (*Newbould & Harley* 4331). The leaves also are used for cleaning copper instruments (*Abeid* 147).
    There is a report of this species being a host in Kenya of whitefly, a pest of tapioca (*Robertson* 3255)
CONSERVATION NOTES. A widespread and common weedy species, here assessed as of Least Concern for conservation.
NOTES. *Waltheria indica* and *W. americana* were published simultaneously. R. Brown (in Narrat. Exped. Congo, App. [Tuckey] 5: 478, 484 (1818)) apparently was the first to unequivocally combine the two names, and he adopted *W. indica* for the combined species. When the two names are considered synonyms, Brown's choice must be adopted (see also Greuter et al., Regnum Veg. 138: 22; Art. 11, Ex. 18. (2000)).

The genus *Waltheria* is poorly represented in Africa. *Waltheria indica* presumably is American in origin and given the great variability in African and Malagasy populations probably represents multiple introductions. The only indigenous African species is *W. lanceolata* Mast., which is confined to West Africa. It can be readily distinguished from *W. indica* by its ovate (versus linear) floral bracts. The recently described *W. laxa* Thulin is based on teratological material of *W. indica*. Strikingly similar inflorescence and floral abnormalities were described and illustrated by Scott (in Bothallia 12: 452, f. 16 (1978)) and noted by Verdoorn (1981), the former citing material from South Africa and Zimbabwe and including at least one specimen that had separate branches with teratological and normal inflorescences and flowers.

INDEX TO STERCULIACEAE

New names validated in this part

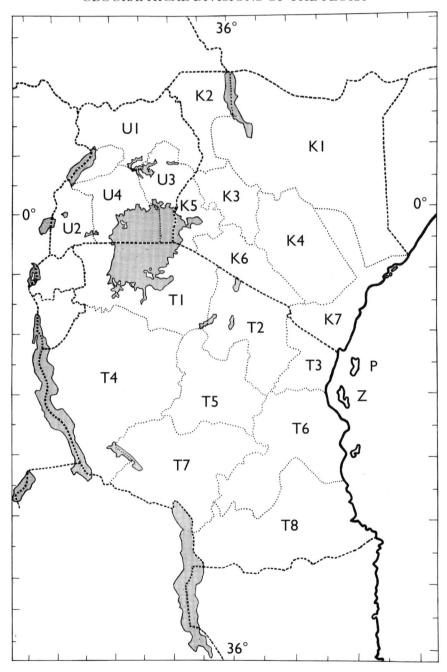

PLANTS PEOPLE
POSSIBILITIES

First published in 2007 by
Royal Botanic Gardens, Kew
Richmond, Surrey, TW9 3AB, UK
www.kew.org

ISBN 978 1 84246 185 3

British Library Cataloguing in Publication Data
A catalogue record for this book is available from the British Library

Design and typesetting by Margaret Newman,
Kew Publishing, Royal Botanic Gardens, Kew.

Printed in the UK by Hobbs the Printers

For information or to purchase all Kew titles please visit
www.kewbooks.com or email publishing@kew.org

All proceeds go to support Kew's work in saving the world's plants for life